5 Steps to Surviving Chemistry:
Tips for Understanding a Challenging Course

By Julie C. Gilbert

Illustrated by Tim Sparvero

Aletheia Pyralis Publishers

For information about special discounts available for bulk purchases, sales promotions, fund-raising and educational needs, please email: juliecgilbert5steps@gmail.com.

http://www.juliecgilbert.com/
https://sites.google.com/view/juliecgilbert-writer/

Not ready to dive into chemistry stuff?

Love Science Fiction or Mystery?

Choose your adventure!
Visit: http://www.juliecgilbert.com/

For details on getting free ebooks

Endorsement:

"Excellent coverage of key topics for chemistry students of all levels (high school and college)"
~ David A. Hunt, Ph.D., The College of New Jersey, Professor of Chemistry

Dedication:

To the teachers and mentors who inspired me as a student.
To the students who inspire me as a teacher.

Special Thanks:
Tim Sparvero for the excellent illustrations.
http://www.infinityroads.com/commission-me/
E. Roy Worley for lending his voice to the narrated version.

Dr. David Hunt, The College of New Jersey
Dr. Rita King, The College of New Jersey

Table of Contents:

Introduction:

Dear Frustrated Chemistry Student (and his/her parent or guardian or "friend"),

You may be wondering …

Who is this book for and how do I use it?

The book is aimed at first-year, college prep chemistry students. I'm not saying that it can't help someone in a lower level or a higher level, first-year course or even the poor soul in college. But the bulk of my experience comes from working with high school CP classes. While you can skip around as necessary, I recommend finishing this introduction.

You can refer to your textbook or an internet search engine for formal definitions of any term that appears in here, but mostly, I'm going to be paraphrasing and providing translated definitions where possible.

What this book is not?

It's not a "guide for dummies." Chemistry is hard. If you survey adults about the toughest course they took in high school or college, many will say Chemistry. It's practically got its own alphabet and language.

Why is this book important?

Chemistry textbooks are great. They're also often written in a manner that's difficult for some students to grasp.

Chemistry teachers are great too. But they have a lot of material to cover in a little time, so sometimes the pace is faster than a student can handle with class-time alone. They're also human, at least for now. They can speak fast, write weird, look strange, or find some other way to simply not connect well with you as a student. Keep in mind, what one student hates another will love.

I've seen several chemistry guides aimed at prepping the high fliers for the AP Chemistry exam or the SAT II exam, but there aren't too many books meant to walk one through a normal, college prep course.

What is Chemistry and why do I have to study it?

Short Answer:

Because the powers that be said so and/or it's in your course of study to graduate.

Longer Answer:

Chemistry is the study of matter and energy. It's called the "central science." One of my old college professors said biology was basically chemistry applied to living things. If you get into higher level physics, there are a lot of parallels to chemistry. Science is broken down into many subjects, but at some point, each one is going to have some chemistry inside it.

Quick Examples:

Forensic Science: Science as it applies to the law and cool things like solving crime. If you want to know the components of a bomb, you'd best know your chemistry. If you want to match a criminal's DNA to a sample found at the crime scene, you should understand the chemistry behind that biology.

Physics: Quantum theory, light, and energy are just a few of the places where physics and chemistry overlap.

Biology: Everything that's alive is made up of elements. So, at its source, it's all chemistry. That's a horrifying thought to many high school students.

Environmental Science: Having an understanding of chemistry can help with topics that affect the environment. The effects of acid rain on weather patterns and alternate energy sources are two topics that you can approach better if you get the chemistry behind them.

Bonus Application: Cooking and baking, food science in general, are pretty much applications of chemistry.

5 Steps to Surviving Chemistry:

Step #1: Have a helpful mindset and a definition of success
Step #2: Know how to use units to your advantage
Step #3: Know how to manipulate equations
Step #4: Know which terms matter and how to apply those terms
Step #5: Know when and how to ask questions and seek help

5 Steps to Surviving Chemistry: (The annotated version ...)

Step #1: Have a helpful mindset and a definition of success
What is success to you? An A? a B? a passing grade? Not getting killed by your parents? It's going to sound lame to start out with "just believe you can do it" but there's a small grain of truth in there. If your mind is completely consumed with thoughts like "I'm so lost. This sucks. When am I ever going to use this!" there's very little room for actually learning the topic. Chemistry like most things is something you can get better at with practice.

Step #2: Know how to use units to your advantage
Units are your friends, and crossing out those suckers can be

highly satisfying. Almost anything that can be done with a proportion can be done with dimensional analysis. That's a fancy way of saying make those units work for you.

Step #3: Know how to manipulate equations
Many topics have at least one math equation. Solving for one of the variables will be important.

Step #4: Know which terms matter and how to apply those terms
There's usually a vocabulary set to understand before any topic can be discussed intelligently. "I'm just no good at memorizing" can be true, but it's usually not. I've had students who failed element symbols quizzes, but could perform in two-hour plays. Ask the same student who claims not to be able to memorize what the stats are for their favorite baseball pitcher and prepare for a lecture. Another former student struggled in class and then quoted what year my car had to be because after that year there was a subtle change to how the company designed the trunk. If you're in a college prep chemistry class, you've convinced somebody you're capable of handling the workload.

Step #5: Know when and how to ask questions and seek help
"I don't get it" is an okay starting point, but you quickly need to move beyond that into specifics. What's the best way to figure out where you got lost? Who can you turn to for help? Friends, tutors, and teachers are all options, so I'll discuss each briefly.

Embrace the weird and the wonderful:

If you have to take chemistry, you might as well try to enjoy some of the class. You may think your teacher's specially trained in boring you, but the subject itself has a lot to offer. Don't be afraid to look further into something that looks interesting.

Conclusion:

One of my colleagues maintains that "chem is try." While most students groan at that, there's some truth to it. Life's not fair. There will be kids in your class who don't seem to pay attention, don't put in much effort, and still ace every test. Most, however, will struggle with the course to some degree.

The chapters will be organized by topic, so use them as you need.

Still not convinced?
Go here to see the free sample chapters.
https://sites.google.com/view/juliecgilbert-writer/home
(It's under the nonfiction tab.)

Note: Curriculum—that's the school's grand plan for what you should learn and when you should learn it—changes every couple of years. As the Next Generation Science Standards become more widely adopted, you might see less and less of this exact layout, but for now, it's still relevant. The foundations of the subject haven't changed for a few hundred years. No matter what bells and whistles will be used to present the subject in the future, the principles presented here will help. Many chemistry textbooks have the chapters laid out in this general order. The textbook also contains much more information. This book is only meant to hit the major topics that many first-year courses touch upon.

I'll try to update my Google site with sample worksheets and keys if you want to practice things in more depth. You can always email me at juliecgilbert5steps@gmail.com if you want something specific or have questions. Please put "Chemistry" in the subject line.

Chapter 1:
Dimensional Analysis and Significant Figures

Introduction: The Ally and the Enemy

Chemistry has a lot of formulas that allow us to explore the topics in more depth. Any time you wrestle with formulas or measurements, you're going to be dealing with units. Despite popular belief, units are your friends. They will make your life easier if you learn to let them do their job. If you fight them, they will haunt your dreams.

Note: As time has become a rare commodity, many schools have phased out the discussion about significant figures. Even I've taken to just telling my students how many significant digits they should go to, instead of waiting for them to figure out how many they should be able to get from a particular instrument. As you wade through this course—and much of life for that matter—you have to pick which battles you want to fight.

Significant Figures and Scientific Notation:

Scientific Notation: a way of presenting massively large or extremely small numbers to something reasonably readable. Scientific notation follows the pattern $M \times 10^x$ (your M value must be greater than 1 but less than 10).

Significant Figures: Besides being pure evil, sig figs are digits that carry meaning. In other words, it's all the digits you know for sure and one digit that you're guessing at. They're based off of measurements from actual instruments, whether they're given to you or you're asked to measure something yourself.

For example, if you read a graduated cylinder (the tall, skinny glass thing your teacher wants you to grab to measure liquids), that has dash marks every milliliter (mL) you should be able to get one decimal place. This is because you can be certain up to the milliliter and then have to estimate (guess at) the last digit, the tenth of a milliliter.

In general, significant figures are based on what can be definitely measured plus one digit. So, for example, if you have a metric ruler you could see if something is 2.24 centimeters. That last digit is estimated though because there is no definite line for you to measure and see for sure. So, one person may look at the metric ruler and say something is 2.27 cm and another say 2.26 cm. I would say, in most cases, it doesn't matter which you choose, so long as the rest of your measurements follow consistently.

Significant figure rules can be found all over the place:
http://chemistry.bd.psu.edu/jircitano/sigfigs.html
Here's a summary:
Non-zero digits = significant
Zeroes sandwiched between non-zero digits = significant
Zeroes following a decimal point and a real digit = significant. (For example, .0002 has one significant figure, but 1.0002 has five significant figures.)

Trailing zeroes are not significant unless there is a decimal point present after. (For example, 2000 has one significant figure, but 2000. has four significant figures.)

Examples:
2,232,000,000 is only 4 sig figs
43,432,000,003 is 11 sig figs
3443.00 contains 6 sig figs (By putting in the 0's after the decimal point, I am deeming them significant. The assumption is that whatever instrument this number was derived from could be estimated to the hundredths place—that second 0.)

Note: the official abbreviation for significant figures is SF, not sig figs.

Why use scientific notation?
If I wanted to indicate the first number 2,232,000,000 to 5 sig figs, I could simply draw a line over the top of that first 0, the fifth digit if you're counting from the left side. Since this is difficult to show while typing, I will put it in scientific notation instead.

2,232,000,000 to 5 sig figs = 2.2320×10^9

Note: When doing problems throughout the course (density, Molarity, etc), you want to show your final answer to the least number of sig figs. Remember that significant figures are derived from a measurement of some sort.

Common Mistakes in Rewriting #'s to the Correct Significant Figures:

- Failing to preserve the place values (10's spots). This will change the number. For example, 235,423 rewritten to three sig figs is 235,000 not 235. If somebody owed you $235,000 and decided to give you $235, you would be pretty upset.

- Not rounding correctly. Look to the digit immediately right of the last SF you're allowed to keep and round. 5 and up round up, 4 and down round down, leave alone. For example, 2,415,487 written to 3 sig figs is 2,420,000 because the 5 will round up.

Mathematical Operations:
Addition and Subtraction: First, line up the decimal points. If I wanted to add 22.46 + .2569, I would need to line up the decimal points, do the addition, and then round to two decimal places, since we are assuming that the 6 in the first number is already an estimated digit.

22.46
+ .2569
22.7169 → 22.72

Multiplication in Scientific Notation:
Multiply the M values and add the exponents. Then, adjust for significant figures and your new M value. For example, if your problem is (2.5×10^2) (5.0×10^4), you would take 2.5 x 4 to get your new M and then add 2+4 to get the new exponent.

2.5×10^2
$\times\ 5.0 \times 10^4$
$12.5 \times 10^6 \rightarrow 1.25 \times 10^7 \rightarrow 1.3 \times 10^7$

One part that can get tricky is if you have a negative exponent, then you may be adding a negative number (in essence subtracting).

Division in Scientific Notation:
General form for a number in scientific notation: $M \times 10^x$
Divide the M values and subtract the exponents (x). Then, adjust your new M value to the proper number of sig figs.

Common Mistakes in Scientific Notation Calculations:

1. Addition/Subtraction: You must make the exponents the same. You do NOT add/subtract the exponents.
2. Multiplication/Division: You must add exponents in multiplication problems and subtract exponents in division problems. Please watch out for negative signs in exponents.

Dimensional Analysis (Conversion Factor Method):

Conversion Factor: a way of presenting a particular equivalent value. For example, 1 dollar = 4 quarters. Think of it as an equivalency expressed as a fraction. That might sound very fancy, but it simply means that the numerator and the denominator represent the same quantity in different units.

Dimensional Analysis: You might hear it called fence-posting. This is a tool that allows you to change from one unit to another. It's just a way of presenting stuff that makes dealing with the units easier. You actually deal with equivalent values all the time. Ever bought something with all quarters? You instinctively used dimensional analysis in your head.

Units: the good guys. Units are a measure of the magnitude of whatever you're talking about. They will help you identify the information you're given. One of my worksheets has candy conversions on it with strange equivalencies like 4 spearmints = 9 gummy worms. Here, spearmints and gummy worms are units. If I have a certain number of spearmints, I can find out how many gummy worms would be equal to that amount. The units you use in class will depend on the instrument you're dealing with.

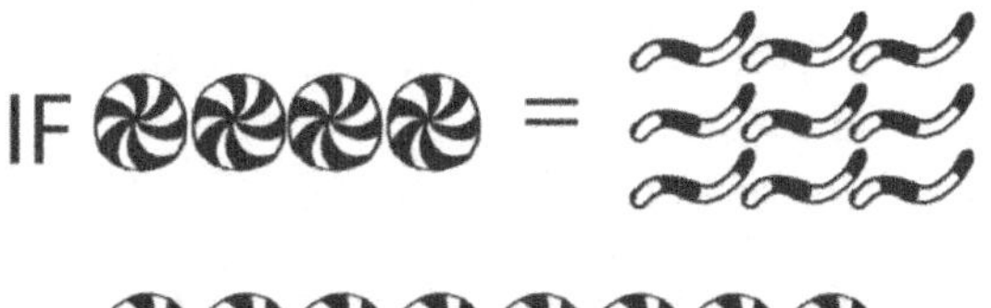

THEN = ?

Common units:

Volume: milliliter (mL), liter (L)
Mass: gram (g), kilogram (kg)
Temperature: degrees Celsius (°C), Kelvin (K)
Time: second (s), minute (min)

What's the point?

Units will help you navigate many of the formulas because they will help you identify the information you're given. The easiest way to do some of the harder chemistry problems is to get good at the conversion factor method.

Steps to Dimensional Analysis:

Step #1: write down the number and unit you would like to convert (starting point)

Step #2: write a multiplication sign followed by the conversion factor you wish to use.

What you go **from** should go on the bottom (in the denominator of the conversion factor). This is the unit you are converting out of.

What you go **to** should go on the top (in the numerator of the conversion factor). This is the unit you are converting to.

Step #3: repeat step 2 as necessary. It might take many conversions to get to the final unit.

Step #4: once you have obtained the unit you want, solve. Numbers on the top (in the numerator) are multiplied. Numbers on the bottom (in the denominator) are divided.

This might sound confusing, so let's see some examples.

Example 1: Let's just say the current Euros to dollars conversion rate is 1 Euro = $1.12. If I want to buy something that costs 8.00 Euros, I could set up a conversion factor to get from euros into dollars, so I would know how many dollars to pay.

8.00 Euros x $\frac{\$1.12}{1\text{ Euro}}$ = $8.96

Explanation: The unit I wanted to change from (Euro) was put on the bottom so the unit would cancel with my original unit. The unit I am left with ($) is on top, since after cancellation it is the only unit left.

Example 2: If I want to buy something that costs $2.50 with quarters, how many quarters do I need? (equivalent value: $1 = 4 quarters)

$2.50 x $\frac{4\ quarters}{\$1}$ = 10 quarters

Explanation: The unit I wanted to change out of, $, goes on the bottom (in the denominator). The unit I wanted to change to, quarters, goes on top (in the numerator). To solve, I end up multiplying by 4. My final unit is quarters.

Example 3: If I have 24 spearmints, how many gummy worms could I get with them? Remember the conversion factor presented above: 4 spearmints = 9 gummy worms.

24 spearmints x $\frac{9\ gummy\ worms}{4\ spearmints}$ = 54 gummy worms

It's like a bartering system. If you have excess French fries and your friend has a pack of cookies he doesn't want, you could work out a price of a certain number of fries needed to buy a cookie. If a cookie costs you 8 fries, the equivalency would be written as 8 fries = 1 cookie. You could then do conversions either starting with fries and ending up in cookies or vice versa.

Example 4: How many cookies could be bought with 40 fries? (equivalent value: 8 fries = 1 cookie)

40 fries x $\frac{1\ cookie}{8\ fries}$ = 5 cookies

Note: You can string as many conversions as you need to together as long as you have the appropriate conversion factors (equivalencies).

The conversion factor method is useful in Chemistry because you can perform calculations involving atoms/ molecules, moles, and mass.

Resources:
More in depth on the zero rules:
http://www.chemteam.info/SigFigs/SigFigRules.html
More on scientific notation:
http://www.edinformatics.com/math_science/scinot.html
More on dimensional analysis:
https://www.chem.tamu.edu/class/fyp/mathrev/mr-da.html

Common Mistakes in Dimensional Analysis:

1. Not starting with the given. You MUST start with the given piece of information.
2. Not matching the units. If you want to cancel a unit, you need to have it on the bottom of the conversion factor. If you want to change to a unit, you need it on the top of the conversion factor.

3. Improper significant figures. The number of significant digits in your final answer should be determined by the starting number given to you not the conversion factors.

Density:

Density often shows up in the early chapters of a chemistry course. It's a relatively simple formula. Students have seen it before. It's great to talk about while discussing units of measurement and introducing laboratory equipment.

Density = $\frac{mass}{volume}$

Density can have several different units, depending on what you use for the mass and the volume. The most common unit for density in many classes is g/mL because those are typically the units used in a lab.

Common Mistakes in Density Problems:

- Not writing down the correct formula.
- Not changing the formula correctly if solving for mass or volume.
- Forgetting to pay attention to sig figs.

Chapter 2:
States of matter and Atoms, Elements, Ions

Introduction: The Puzzle Pieces

If you want to build a puzzle, you need all the pieces. If you want to grasp higher chemistry topics, you need a working knowledge of the building blocks of life and the ways they combine.

Definitions:

Substance: stuff; it consists of physical, tangible matter

Solid (s): a state of matter where the substance in question has a definite volume and definite shape

Liquid (l): a state of matter where the substance flows; it has definite volume but no definite shape

Gas (g): a state of matter where the substance has no definite volume or shape

Physical property: these can be observed (seen) or measured or changed without altering the identity of the substance. Examples include: melting point, boiling point, density, texture, color

- **Intensive physical property:** does not depend on the amount of matter present; for example, density, color, boiling point
- **Extensive physical property:** the value changes depending on the amount of matter present; for example, mass and volume

Chemical property: a property that can be observed (seen) when a substance goes through a chemical reaction. Examples include: toxicity, reactivity, flammability, chemical stability
Physical change: the substance changes state of matter (s, l, g)
Chemical change: the substance combines with another substance (or breaks down) into something new

Physical vs. Chemical Changes:
One of the first things you will likely be asked to do is distinguish between (tell the difference) physical and chemical changes. There are plenty of worksheets online you can use to practice. The key question to ask yourself when trying to make that decision is: does the change alter the substance in some fundamental way? Can I get the original substance back easily without a chemical reaction?

Example:
If I rip a piece of paper in half, I've caused a physical change. I have not changed the paper itself. It's still paper, even if it is in two pieces.
If I burn a piece of paper, I've altered it in a way that cannot be undone. It's no longer paper. It's soot and ash.

Element: the simplest type of substance; made up of atoms with the same number of protons (we'll get to subatomic particles momentarily); these can be found on the periodic table of elements (the single most useful tool at your disposal)
Compound: is formed when two or more elements have been chemically combined; the new substance has its own chemical properties; for example, sodium chloride (table salt; $NaCl$)

Mixture: is formed when two or more substances are physically combined; each part retains its chemical properties

- **Homogenous mixture:** uniform throughout; if you look at a sample from one part, it should look and be the same as a sample from a different part
- **Heterogeneous mixture:** not uniform throughout; if you look at a sample from one part, it could be different than a sample from a different part

Source: www.thoughtco.com/examples-of-mixtures-608353

Atomic Theory:

Matter (stuff) is composed of atoms. Mankind has moved up from Democritus's idea that matter is made up of tiny particles of smaller stuff until you can't break it down any further to John Dalton's theory which gets a little more specific to the modern atomic theory.

Statements of Dalton's Theory:

- elements are made of atoms
- atoms of the same element all have the same mass
- atoms of different elements have different masses
- atoms combine in ratios of small whole numbers (you can't have half a sodium atom)

Source: www.chemteam.info/AtomicStructure/Dalton.html

Modern Atomic Theory:
There are hundreds of thousands of pages on this topic, including side trails into every scientist who made a significant contribution to the discovery of the atom. The models have come a long way. Currently, we believe atoms are protons and neutrons surrounded by electron clouds. That's an overly simplified statement, but general chemistry courses rarely dive into the quantum theory right in the beginning. There will be a chapter dedicated to those concepts shortly.

A general chemistry course is more likely to detail the history of the atom and the discoveries and experiments that led to knowledge of the subatomic particles.

A short history of the atom: A who's who and what's what ...
Democritus: coined the term atom (technically, it was "atomos" Greek for "undivided"). Aristotle, who had more intellectual clout, rejected the idea, so it got buried for a few thousand years. That, ladies and gents, is an early lesson in the dangers of being swayed by the popular ideas.
Dalton: came up with the first atomic theory
J.J. Thomson: discovered the existence of electrons

Thomson's model (The Plum Pudding Model):

He basically said that the atom is a positive matrix with areas of electrons scattered throughout, like raisins in plum pudding. Since the idea of plum pudding is foreign to the average American students, picture it like a chocolate chip cookie. The cookie part is the positive matrix and the chips are the negative part. It's a decent theory given what Thomson knew at the time. It's wrong, but it was a decent start and challenged others to prove or disprove it. That's what science is supposed to do. Incidentally, Thomson called these areas of negative charges "corpuscles." I for one am grateful that creepy name didn't stick.

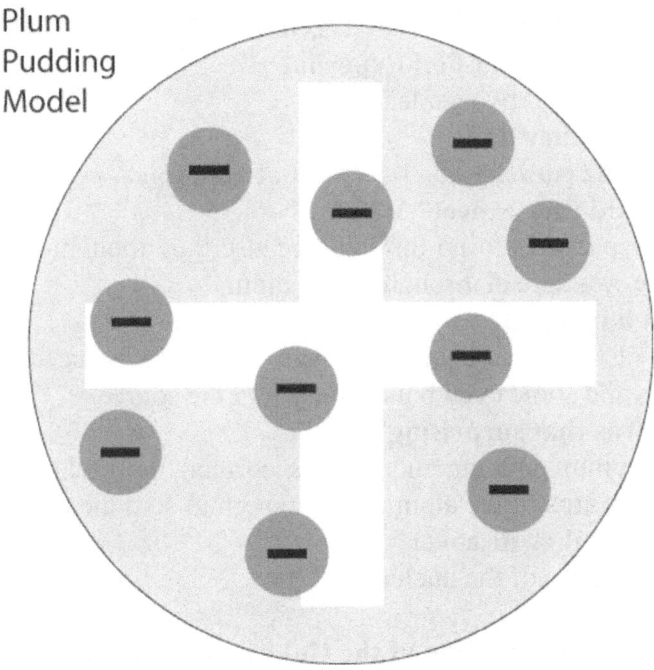

The Gold Foil Experiment: (the one that took out Thomson's model)

The Gold Foil Experiment is one of the first ones chemistry teachers will gush about. It was run by Ernest Rutherford and two of his students, Hans Geiger and Ernest Marsden. They shot alpha particles (helium nuclei which are small and positively charged) at a thin sheet of gold foil. Since the working theory of the day was Thomson's plum pudding model, they expected all the particles to go straight through the gold foil. Most of the alpha particles indeed went through, but a few bounced at weird angles and some even bounced back at the source. This led them to conclude that the area of positive charge was not spread throughout the atom as the plum pudding model suggested but rather confined to a very small area of the atom. This is the nucleus.

Let me state all that again point-by-point:

Who ran the Gold Foil Experiment?
Rutherford and two students
What did they do?
Shot alpha particles at a thin sheet of gold foil
What did they expect? Why?
All the particles to go through because they thought the positive charge was spread throughout the atom
What happened?
Most of the particles went through, but some bounced at strange angles and some even bounced back at the source.
Why was that surprising?
If the plum pudding model was correct, there shouldn't have been any area of the atom dense enough to stop the particles.
What did they discover?
The existence of the nucleus

Figure: an illustration of the Gold Foil Experiment

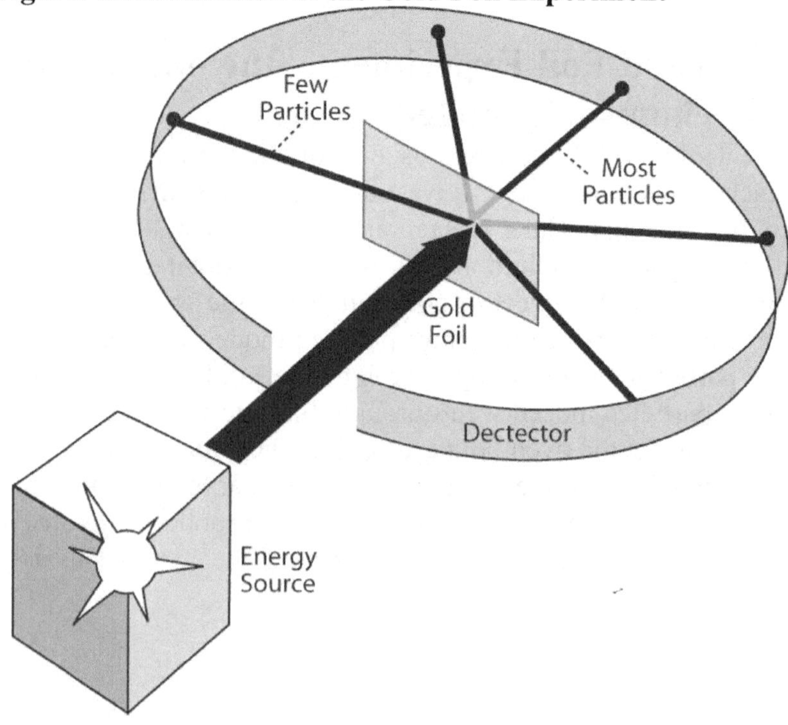

Just in case you want to read more:
www.ck12.org/book/CK-12-Physical-Science-For-Middle-School/section/5.2/
http://myweb.usf.edu/~mhight/goldfoil.html

Structure of the Atom: (yet more definitions)

Protons (p): positively charged subatomic particles located in the nucleus

Neutrons (n): subatomic particles located in the nucleus that add mass to an atom but do not carry a charge

Electrons (e^-): subatomic particles located outside the nucleus that does not add much mass but does carry a negative charge; in a neutral atom, this will be the same number as the number of protons

Atomic number: the number of protons in an atom (positive charge)

Mass number: the number of protons plus the number of neutrons (mass # = p+n)

Isotopes: atoms of the same element with a different mass due to varying numbers of neutrons

Two types of notation for isotopes:

Hyphen notation: shows only the element name or symbol, a hyphen, and the mass number; for example, hydrogen-3

Nuclear symbol: shows the element symbol with a left superscript of the mass number and a left subscript atomic number

Example: What is the hyphen notation for a nuclide with 13 protons and 11 neutrons?

Steps to solving the problem:
1. Look up element with 13 protons (aluminum)
2. Add the number of neutrons and protons (13+11 = 23)
3. Write your answer in the hyphen notation: aluminum–23

Tip: Learn the following Element Names and Symbols:

It will make your life a whole lot easier if you can recognize certain elements by their names and symbols. Which elements and how many will vary by class. Here's the list I give my students at the beginning of the year.

Note: It might differ for your class, but in my class, spelling counts, especially because I give them a periodic table with the names spelled out completely. Pay extra attention to spelling of beryllium, fluorine, phosphorus, nickel. Also, be careful with the symbols Mg and Mn. I'm using the American spelling for phosphorus and sulfur, but if you're not in the United States, you might want to check which spelling you should use.

Important Element Names and Symbols:

#	**Symbol**	**Name**	#	**Symbol**	Name
1	H	Hydrogen	25	Mn	Manganese
2	He	Helium	26	Fe	Iron
3	Li	Lithium	27	Co	Cobalt
4	Be	Beryllium	28	Ni	Nickel
5	B	Boron	29	Cu	Copper
6	C	Carbon	30	Zn	Zinc
7	N	Nitrogen	31	Ga	Gallium
8	O	Oxygen	32	Ge	Germanium
9	F	Fluorine	33	As	Arsenic
10	Ne	Neon	34	Se	Selenium
11	Na	Sodium	35	Br	Bromine
12	Mg	Magnesium	36	Kr	Krypton
13	Al	Aluminum	37	Rb	Rubidium
14	Si	Silicon	38	Sr	Strontium
15	P	Phosphorus	47	Ag	Silver
16	S	Sulfur	50	Sn	Tin
17	Cl	Chlorine	53	I	Iodine
18	Ar	Argon	54	Xe	Xenon
19	K	Potassium	55	Cs	Cesium
20	Ca	Calcium	56	Ba	Barium
21	Sc	Scandium	79	Au	Gold
22	Ti	Titanium	80	Hg	Mercury
23	V	Vanadium	82	Pb	Lead
24	Cr	Chromium			

Brief Introduction to the Periodic Table of Elements:

Atomic mass: mass of an atom; measured in amu (atomic mass units), which is approximately the number of protons and neutrons in an atom

Average atomic mass: a weighted average of the naturally occurring isotopes of an atom

Concept of weighted average:

I'm pretty sure I just lost about half of you there with the definition of average atomic mass. Imagine you have a bag of mixed marbles. For simplicity, let's say there are only two sizes of marbles. You have a few bigger marbles and a lot of smaller marbles. The bigger marbles might have a greater mass, but their effect on the average will be smaller since there are less of them.

Steps to solving average atomic mass problems:

1. Find percent as a decimal or change percents to decimal percents (divide by 100)
2. Multiply decimal % by the respective weights
3. Add the masses
4. Adjust for proper sig figs if necessary

Example: If 13 marbles weigh 24.3 g, 6 marbles weigh 30.9 g, and 14 marbles weigh 21.2 g, what is the average atomic mass?

Step 1: 13 + 6 + 14 = 33 $\frac{13}{33} = .3939$ $\frac{6}{33} = .1818$ $\frac{14}{33} = .4242$

Step 2: [(.3939)*(24.3 g)] + [(.1818)*(30.9 g)] + [(.4242)*(21.2 g)] =

Step 3: 9.57177 g + 5.61762 g + 8.99304 g = 24.18243 g

Step 4: 24.2 g

Why is atomic mass not a whole number?

It's a weighted average. See above. With that much multiplying, dividing, and adding going on, it's unlikely to be a whole number.

What's the difference between mass number and average atomic mass?
Mass number is the mass of a particular isotope. You get it by adding the number of protons and the number of neutrons.

Atomic mass is a weighted average of the naturally occurring isotopes of an element.

What do you mean by "naturally occurring"?
Technically, there could be a lot of isotopes for every element, but in reality, nature favors certain isotopes because of their stability. So, while it might be possible to get a carbon-6, meaning an isotope without any neutrons, it's very unlikely. There are three naturally occurring isotopes of carbon: C-12, C-13, and C-14.

Quick Periodic Table Orientation Lesson:

Group (family): a column of the periodic table of elements. There are eighteen of these. Groups 1, 2, and 13-18 are considered main group elements. These are the ones we'll study in greater detail to figure out reactivity.

Group 1	Group 2	Group 13	Group 14	Group 15	Group 16	Group 17	Group 18
H							He
Li	Be	B	C	N	O	F	Ne
Na	Mg	Al	Si	P	S	Cl	Ar
K	Ca	Ga	Ge	As	Se	Br	Kr
Rb	Sr	In	Sn	Sb	Te	I	Xe
Cs	Ba	Ti	Pb	Bi	Po	At	Rn
Fr	Ra						

Notable groups with their special names:

- Group 1: alkali metals (except hydrogen)
- Group 2: alkaline earth metals

- Group 17: halogens (very reactive)
- Group 18: noble gases (not very reactive)

Period (series): a row of the periodic table of elements. There are seven of these. Later, you will see that they correspond to the energy levels. It will help you write electron configuration when the time comes.
Metals: located on the left side of the periodic table; note that hydrogen is placed above the first group (the alkali metals) due to the way it reacts, not because it is itself a metal.
Non-metals: located on the right side of the periodic table
Diatomic elements: when found in nature, they will always come in pairs. (The prefix "di" means two.)

What are the diatomic elements?
Nitrogen, oxygen, fluorine, chlorine, bromine, iodine, and hydrogen

How do I remember that?

There are a few different memory hooks to help you memorize the diatomic elements. I prefer this: start at #7 nitrogen, form a seven (across two—oxygen and fluorine—and down three—chlorine, bromine, and iodine), and add hydrogen (the seventh one).

One of my colleagues uses HONClBrIF because it sounds weird. Those are the symbols for hydrogen, oxygen, nitrogen, chlorine, bromine, iodine, and fluorine. If it works for you, go with it.

How can the periodic table tell us the ion charge of an element?

Valence electrons: outer shell electrons. These are the electrons available to be gained, lost, or shared to obtain a stable configuration. Usually, this translates to eight electrons in the outer shell.

How many valence electrons do each of the main groups have?

- Group 1 = 1 valence electron
- Group 2 = 2 valence electrons
- Group 13 = 3 valence electrons
- Group 14 = 4 valence electrons
- Group 15 = 5 valence electrons
- Group 16 = 6 valence electrons
- Group 17 = 7 valence electrons
- Group 18 = 8 valence electrons (except for helium, that one only as two valence electrons)

Octet rule: a guiding generalization that main group elements (1, 2, and 13-18) strive to get eight electrons in their outer shells.
Notable exceptions to the octet rule: hydrogen, helium, lithium, beryllium, boron; they are all content with less than eight electrons in their outer shells.

Electron dot diagram (Lewis dot structures) for elements: Write the element symbol and then write a dot to represent each of the electrons in the valence shell of an atom. I believe there is a formal way you're supposed to add them: top right, bottom, left, top, then around again right, bottom, left, top. I'm not that picky in class as long as the students put no more than two electrons per side.
Ion: what an atom becomes when it gains or loses electrons; also, a group of covalently bonded atoms with an overall net charge, but here, we're referring to atoms that become ions
Cation: a positive ion; formed when an atom loses one or more electrons
Anion: a negative ion; formed when an atom gains one or more electrons

Why do atoms become ions?
The key to this concept is stability. Group 18 elements—the noble gases—have complete outer shells of electrons. For most of these, that means eight electrons in the outer shell, but helium only has two electrons, so automatically, those are in its valence shell. The other elements really want to be like Group 18 elements. They want a stable electron configuration.

What kind of ion does each group form?

- Group 1 elements only have one valence electron, so they form +1 ions by losing that one electron.
- Group 2 elements have two valence electrons, so they form +2 ions by losing both of those electrons.
- Group 13 elements have three valence electrons, so they form +3 ions by losing all three of those electrons.
- Group 14 elements have four valence electrons, so

they can form +4 or -4 ions. However, they tend to share electrons rather than give up or gain electrons.

- Group 15 elements have five valence electrons, so they form -3 ions by gaining three electrons to complete their outer shell.
- Group 16 elements have six valence electrons, so they form -2 ions by gaining two electrons to complete their outer shell.
- Group 17 elements have seven valence electrons, so they form -1 ions by gaining one electron to complete their outer shell.

Chapter 3:
Atomic Properties and Periodic Trends

Introduction:

It's convenient to think of atoms as Niels Bohr did. His model's not perfect, but it nicely illustrates the idea that electrons can't be just anywhere. The energy of each level is quantized (has a certain value).

Definitions:

Photon: a particle of light energy; a small packet of energy
Energy: can exist in many different forms. The energy of a photon can be calculated because it's related to wavelength.
Wave nature of light: light can behave as a wave; it has properties like a wave
Particle nature of light: light can behave as a particle; it has properties that relate to a particle

Source for delving deeper into dual nature of light:
www.nobelprize.org/nobel_prizes/themes/physics/ekspong/

Light Problems:

There are really only two equations you will likely be asked to know or use.

Important Equations:

$E = h \cdot \nu$ $\quad$ $c = \nu \cdot \lambda$

The first is energy equals Planck's Constant times the frequency. The second is the speed of light equals the frequency times the wavelength.

Other important information: 1 mol = 6.02 x 10^{23} atoms/molecules/particles
1 nm = 1 x 10^{-9} m   h = Planck's Constant = 6.626 x 10^{-34} J•s
c = speed of light = 3.00 x $10^{8}\,\frac{m}{s}$

A note about units: inverse seconds (s^{-1}) and hertz (hz) are the same thing.

Bohr Model:

The Bohr model shows an atom as a small, positively charged nucleus surrounded by electrons traveling in circular orbits around it. In order to draw Bohr models, you'll need to know that the first energy shell, represented by a ring, can hold two electrons. The second energy shell, represented by a ring surrounding the first one, can hold eight electrons, and the third can hold eighteen. Not that it comes up regularly in conversation, but I don't know of any chemistry teachers who make their students draw Bohr models above calcium, which is #20.

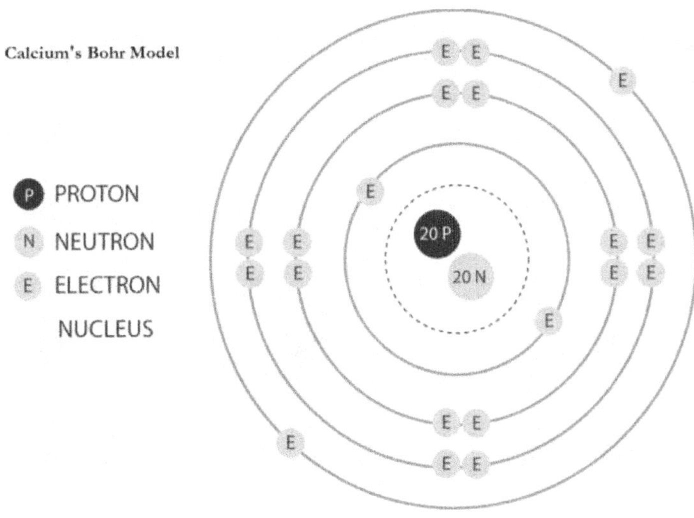

Techniques for drawing Bohr models will vary. To keep it simple, I've taken to having students draw a small circle surrounding whatever number of protons should be in the nucleus. (If I get ambitious, it's a small circle surrounding whatever number of protons and the same number of neutrons in the nucleus.) Around this, I draw a circle then write $2e^-$ on that circle to indicate that the energy level (ring) would be occupied by two electrons. Then, I'd draw another circle and write $8e^-$. If necessary, I'd draw a third circle and place whatever number of electrons there should be on that energy level (ring).

Once again, the model is not perfect, but it is convenient for understanding the next concept.

Here is a link that has a decent view of Bohr models:
https://dashboard.dublinschools.net/lessons/resources/bohr_1404181976_md.gif

What occurs with an electron to result in electromagnetic energy of a certain frequency?

Ground state: when electrons occupy the lowest energy sublevel available

Excited state: when one or more electrons occupies an energy level above the ground state

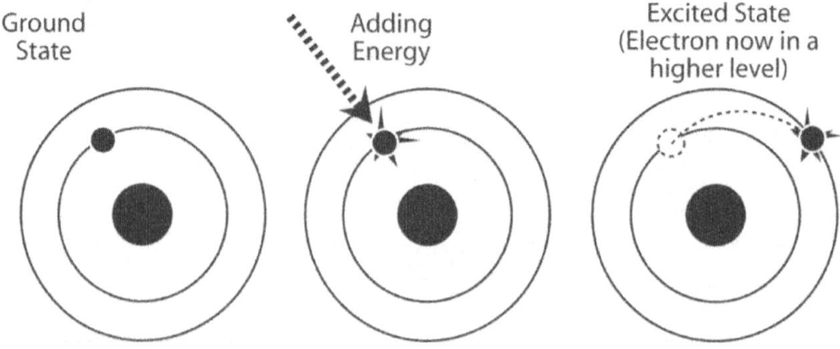

It takes energy to move an electron from a lower energy shell (level, ring) to a higher energy shell. An electron will release that energy if it falls from a higher energy shell to a lower energy shell. The farther an electron falls, the more energy it will release. This energy will manifest itself in the form of light. Because the light has a certain frequency, it will show up as a specific color if it happens to fall within the visible spectrum. Emissions could be in the ultraviolet or infrared portions of the electromagnetic spectrum as well. These can be measured and recorded. The pattern of these spectral lines is unique to each element. This is what's meant when you hear that emissions are unique to each element and that they form an emission spectrum.

For example: Hydrogen can emit light with a wavelength of 656.2 nm (red), 486.1 (blue-green), 434.0 (blue-violet), and 410.1 (violet). Hydrogen can only emit visible light at those wavelengths. This means that the pattern of those colors can let us identify hydrogen by its spectral lines. This is true for other elements too. Because the spectral line patterns are unique to each element, they can function as a fingerprint of sorts to let us identify them.

Note: Hydrogen contains only one electron, so how can it emit light at multiple wavelengths? The answer is surprisingly simple. There are many hydrogen atoms in any given sample. In fact, there are probably millions and millions of atoms in any given sample. We'll get a firmer grip on just how tiny atoms are in the moles chapter, but for now, accept that they're very tiny.

Source:
http://chemed.chem.purdue.edu/genchem/topicreview/bp/ch6/bohr.html

Electron Configuration and Quantum Numbers:

Address analogy:

Take the analogy of a typical address. A person living in the United States lives in a particular state, on a certain street, in a numbered house or apartment building, and they have a name. The same sort of holds true for electrons. Quantum numbers are simply the address for each electron. Each electron will have four quantum numbers.

The principle quantum number (n) indicates which main energy level that electron resides in (state). The angular momentum quantum (l) number indicates the shape of the orbital (street). The magnetic quantum number (m) indicates the orientation of that orbital around the nucleus (house), and the spin quantum number (+½ or -½) tells which way the electron is moving (name).

Each piece of information is slightly more specific, but any one of them alone basically means nothing. Saying a particular electron is in energy level 3 is akin to saying a particular person lives in Georgia. It's not really useful because a whole lot of people live in the state of Georgia. If I tried to send a letter to that person by simply addressing it by their name and big bold letters saying GEORGIA, it would be returned to me.

Orbital: a place where you're likely to find an electron. Each orbital can hold up to 2 electrons.

What do all those numbers and letters mean?

The principal quantum number (n) – main energy levels

The period number on the Periodic Table of Elements corresponds to the highest occupied energy level for an element with its electrons in the ground state (lowest possible). They can be 1, 2, 3, 4, 5, 6, or 7. We haven't discovered or successfully synthesized elements with electrons in energy level 8 while in the ground state. (Electrons in the 8th level would get there by being excited. This requires an input of energy.)

Angular momentum quantum number (*l*) – shape and Magnetic quantum number (m)

l can be 0 but the highest it can be is (n-1). So, if n = 1, then the angular momentum quantum number possibilities are limited to 0. Thus, the first energy level only contains an s orbital.

- 0: s, spherical, these can hold only one orbital containing up to two electrons
- 1: p, lobed shape; there can be up to 3 p orbitals in any given main energy level except the first.
  Therefore, if there are 3 p orbitals that can each hold 2 electrons, there can be a total of 6 electrons held in a p subshell.
- 2: d, flower shaped; the d subshell contains 5 orbitals, therefore, d subshells can hold 10 electrons
- 3: f, complex; the f subshell contains 7 orbitals, therefore, they can hold 14 electrons

The magnetic quantum number is determined by the angular momentum quantum number. The range for *l* is –*l* to +*l*. It shows the number of orbitals and their orientation within a subshell.

Here is a link to a good summary of angular momentum quantum number:
http://study.com/cimages/multimages/16/angularmomentumshapes2.png

Most high school curriculums will barely brush upon this part of the topic, but if you would like to dive much deeper with quantum numbers go here:
https://chem.libretexts.org/Core/Physical_and_Theoretical_Chemistry/Quantum_Mechanics/10%3A_Multi-electron_Atoms/Quantum_Numbers

Spin quantum number (s)
Within an orbital, each electron can spin clockwise or counterclockwise. +½ or -½ simply means that the two electrons in any one orbital must be spinning in opposite directions.

For example: $1s^2$ is the electron configuration for helium, which has two electrons. Both electrons are in main energy level 1 (principal), in a spherical orbital (angular momentum = 0), and one is spinning clockwise and the other counterclockwise.

3 Rules for Electron Configurations:

1. Aufbau Principle
2. Hund's Rule
3. Pauli's Exclusion Principle

Aufbau principle: electrons go into the lowest energy orbital available

Hund's rule: orbitals of equal energy each receive one electron before any receives a second electron

Pauli's exclusion principle: no two electrons can have the same four quantum numbers. This essentially means that two electrons in the same orbital will differ in their direction of spin.

What kinds of electron configurations will I have to know?

There are several kinds of notation, but full electron configuration is generally the middle ground one in terms of effort to write down. Noble gas notation is the easiest in terms of the amount of writing you have to do, and orbital notation is arguably the hardest. To understand noble gas notation, you need a working knowledge of full electron configuration.

Trick: Take a printable periodic table and label groups 1 and 2 (that is the first two columns) as the "s-block." Label groups 13-

18 the “p-block.” Label groups 3-12 as the “d-block.” (The “f-block” is at the bottom of the periodic table, but generally speaking, I ignore that block because I think you can get a good grasp of the concept without going that high. But I expect some of you will need the information at least in noble gas notation.)

All elements that fall in the “s-block” have electron configurations that end with s. All elements that fall in the “p-block” have electron configurations that end with p. All elements that fall in the “d-block” have electron configurations that end with d. And all elements that fall in the “f-block” have electron configurations that end with f.

Once you familiarize yourself with the blocks, you can always figure out any configuration.

Note: Helium is part of the “s-block” for the purposes of electron configuration.

What order do the subshells fill in?
1s, 2s, 2p, 3s, 3p, 4s, 3d, 4p, 5s, 4d, 5p, 6s, 4f, 5d, 6p, 7s, 5f, 6d, etc.

However, it might be easier to see it in a chart form:
http://users.stlcc.edu/gkrishnan/fillingOrder_2.gif

Note: When reading something like this, you need to follow from the end of the arrow to the tip then go to the back of the next arrow and follow it down.

Full electron configuration:

Let's have a look at the first 30 to get a feel for the pattern.

1 H	$1s^1$
2 He	$1s^2$
3 Li	$1s^2\ 2s^1$
4 Be	$1s^2\ 2s^2$
5 B	$1s^2\ 2s^2\ 2p^1$
6 C	$1s^2\ 2s^2\ 2p^2$
7 N	$1s^2\ 2s^2\ 2p^3$
8 O	$1s^2\ 2s^2\ 2p^4$
9 F	$1s^2\ 2s^2\ 2p^5$
10 Ne	$1s^2\ 2s^2\ 2p^6$
11 Na	$1s^2\ 2s^2\ 2p^6\ 3s^1$
12 Mg	$1s^2\ 2s^2\ 2p^6\ 3s^2$
13 Al	$1s^2\ 2s^2\ 2p^6\ 3s^2\ 3p^1$
14 Si	$1s^2\ 2s^2\ 2p^6\ 3s^2\ 3p^2$
15 P	$1s^2\ 2s^2\ 2p^6\ 3s^2\ 3p^3$
16 S	$1s^2\ 2s^2\ 2p^6\ 3s^2\ 3p^4$
17 Cl	$1s^2\ 2s^2\ 2p^6\ 3s^2\ 3p^5$
18 Ar	$1s^2\ 2s^2\ 2p^6\ 3s^2\ 3p^6$
19 K	$1s^2\ 2s^2\ 2p^6\ 3s^2\ 3p^6\ 4s^1$
20 Ca	$1s^2\ 2s^2\ 2p^6\ 3s^2\ 3p^6\ 4s^2$
21 Sc	$1s^2\ 2s^2\ 2p^6\ 3s^2\ 3p^6\ 4s^2\ 3d^1$
22 Ti	$1s^2\ 2s^2\ 2p^6\ 3s^2\ 3p^6\ 4s^2\ 3d^2$
23 V	$1s^2\ 2s^2\ 2p^6\ 3s^2\ 3p^6\ 4s^2\ 3d^3$
24 Cr	$1s^2\ 2s^2\ 2p^6\ 3s^2\ 3p^6\ 4s^1\ 3d^5$
25 Mn	$1s^2\ 2s^2\ 2p^6\ 3s^2\ 3p^6\ 4s^2\ 3d^5$
26 Fe	$1s^2\ 2s^2\ 2p^6\ 3s^2\ 3p^6\ 4s^2\ 3d^6$
27 Co	$1s^2\ 2s^2\ 2p^6\ 3s^2\ 3p^6\ 4s^2\ 3d^7$
28 Ni	$1s^2\ 2s^2\ 2p^6\ 3s^2\ 3p^6\ 4s^2\ 3d^8$
29 Cu	$1s^2\ 2s^2\ 2p^6\ 3s^2\ 3p^6\ 4s^1\ 3d^{10}$
30 Zn	$1s^2\ 2s^2\ 2p^6\ 3s^2\ 3p^6\ 4s^2\ 3d^{10}$

Note: Some of the elements break Aufbau Principle. It's been observed that having a half-filled d subshell or a fully filled d subshell is more stable than having a full s subshell. For instance, there is no $3d^4$ or $3d^9$ configuration. Instead, both chromium (Cr) and copper (Cu) borrow an electron from the 4s subshell to obtain the more stable configuration.

Note: Go back and look at these first thirty elements again. Notice how their configuration ends.

For example:

- Si, which has a configuration of $1s^2\ 2s^2\ 2p^6\ 3s^2\ 3p^2$, is in the "p-block," so it's not surprising that its electron configuration ends with a p.
- Zn, which has a configuration of $1s^2\ 2s^2\ 2p^6\ 3s^2\ 3p^6\ 4s^2\ 3d^{10}$, is in the "d-block," so it's not surprising that its electron configuration ends with a d.

Noble gas notation:
This is just like electron configuration, but it is an abbreviated version that refers back to the nearest noble gas <u>before</u> the one you're trying to draw.

For example: Lithium has the electron configuration $1s^2\ 2s^1$ and a noble gas notation of [He] $2s^1$. Now this may not seem like you saved a lot of work but consider sodium's two configurations. $1s^2\ 2s^2\ 2p^6\ 3s^1$ and [Ne] $3s^1$. Simply putting Neon in brackets is saying that sodium has the same electron configuration as neon except for the $3s^1$ which accounts for the 11th electron.

Here's the full chart of the first 30 elements and their noble gas notations:

1 H	$1s^1$	16 S	[Ne] $3s^2\ 3p^4$
2 He	$1s^2$	17 Cl	[Ne] $3s^2\ 3p^5$
3 Li	[He] $2s^1$	18 Ar	[Ne] $3s^2\ 3p^6$
4 Be	[He] $2s^2$	19 K	[Ar] $4s^1$
5 B	[He] $2s^2\ 2p^1$	20 Ca	[Ar] $4s^2$
6 C	[He] $2s^2\ 2p^2$	21 Sc	[Ar] $4s^2\ 3d^1$
7 N	[He] $2s^2\ 2p^3$	22 Ti	[Ar] $4s^2\ 3d^2$
8 O	[He] $2s^2\ 2p^4$	23 V	[Ar] $4s^2\ 3d^3$
9 F	[He] $2s^2\ 2p^5$	24 Cr	[Ar] $4s^1\ 3d^5$
10 Ne	[He] $2s^2\ 2p^6$	25 Mn	[Ar] $4s^2\ 3d^5$
11 Na	[Ne] $3s^1$	26 Fe	[Ar] $4s^2\ 3d^6$
12 Mg	[Ne] $3s^2$	27 Co	[Ar] $4s^2\ 3d^7$
13 Al	[Ne] $3s^2\ 3p^1$	28 Ni	[Ar] $4s^2\ 3d^8$
14 Si	[Ne] $3s^2\ 3p^2$	29 Cu	[Ar] $4s^1\ 3d^{10}$
15 P	[Ne] $3s^2\ 3p^3$	30 Zn	[Ar] $4s^2\ 3d^{10}$

Tip: If you count the number of electrons, you should be able to tell which element is being shown. The number of electrons should equal the number of protons which is indicated by the atomic number.

For example:
Question: Which element is represented by the following electron configuration? $1s^2\ 2s^2\ 2p^4$
Answer: Oxygen. There are 2 + 2 + 4 electrons in this configuration. The total number of electrons is eight, which corresponds to atomic number eight, which is oxygen.

Trick – Cheat Sheet: $1s^2\ 2s^2\ 2p^6\ 3s^2\ 3p^6\ 4s^2\ 3d^{10}\ 4p^6\ 5s^2$ should get you up through strontium.

Be careful though because there are some funny ones in there, most notably Cr and Cu.

Caution: What I have shown you is the electron configuration in the order that the orbitals fill, but some textbooks show numerical order. Be sure you know what your particular teacher is asking for.

For example: I've seen strontium written:
In order that the orbitals fill: $1s^2\ 2s^2\ 2p^6\ 3s^2\ 3p^6\ 4s^2\ 3d^{10}\ 4p^6\ 5s^2$
In numerical order: $1s^2\ 2s^2\ 2p^6\ 3s^2\ 3p^6\ \underline{3d^{10}\ 4s^2}\ 4p^6\ 5s^2$

Quick Review:
Ion: what an atom becomes when it gains or loses electrons
Valence electrons: outer shell electrons. These are the electrons available to be gained, lost, or shared to obtain a stable configuration.
Octet rule: a guiding generalization that main group elements (1, 2, and 13-18) strive to get eight electrons in their outer shells. Eight electrons in the outer shell is stable.

Side Note on Ions:
Valence electrons will be those on the outer shell. You can tell what kind of ion would form by calculating the difference between the number of outer shell electrons and what it would take to get to one of the Group 18 elements.

If you observe the group numbers and can determine how many electrons must be gained or lost, you can tell which type of ion an element is usually going to form.

Groups with a low number of valence electrons, such as Group 1, Group 2, and Group 13, tend to lose electrons to form stable electron configurations. These become positive ions.

Groups with a high number of valence electrons, such as Group 15, Group 16, and Group 17, tend to gain electrons to form stable electron configurations. These become negative ions.

For example: Look at the electron configuration of fluorine: $1s^2$ $2s^2$ $2p^5$ vs. the electron configuration for neon (the nearest noble gas or Group 18 element) $1s^2$ $2s^2$ $2p^6$. Fluorine only needs one electron to fill the 2p subshell, so it will gain one electron to obtain a stable octet. When F gains one electron, it will become an F^{-1} ion.

For example: Look at the electron configuration for sodium: $1s^2$ $2s^2$ $2p^6$ $3s^1$ vs. the electron configuration for neon $1s^2$ $2s^2$ $2p^6$. Sodium (Na) has one too many electrons to have the configuration of neon. Therefore, sodium tends to give up one electron in order to drop its total number of electrons down from eleven to ten. When this happens, Na becomes Na^{+1} because it has eleven protons and ten electrons.

Why bother?

The more you know about electron configuration, the better a grasp you may have of the elements and how they behave (react).

History of the Periodic Table of Elements:

In 1869, Dimitri Mendeleev put together the first periodic table by arranging the elements known at the time by their atomic masses and arranging them by reactivity. His table was good enough that it actually predicted the existence of three undiscovered elements. Later, Henry Moseley put together a table that was arranged by increasing order of the number of protons in the nucleus instead of atomic mass.

A good resource:
http://www.rsc.org/periodic-table/history/about

Periodicity: when something occurs or repeats at regular intervals
Periodic law: the properties of elements are periodic functions of their atomic numbers. **Translation:** if you put the elements in order by increasing atomic number (# of protons), their properties will repeat regularly.

Periodic Trends:

1. Atomic radius
2. Ionization energy
3. Electron affinity
4. Electronegativity

Bonus items to consider: ion size and successive ionization energy.

Key point: atoms strive to be stable. Later, we'll find out the lengths they go to in order to gain stability, but for now, accept that stability is their ultimate goal. This will help you understand some of the trends.

There are always exceptions, but generally, when talking about a periodic trend, we either mean left to right across a period (horizontal row) or top to bottom down a group (column).

Atomic radius: a measure of the size of an atom. The distance between the nucleus and the edge of the electron cloud. In other words, the distance between the nucleus and the farthest electron. There is no distinct "edge" of the electron cloud as electrons are in constant motion and that motion is not circular, despite the models we use. This is why some textbooks define atomic radius as half the distance between two nuclei.

- **Across a period:** atomic radius tends to get smaller because we have not changed energy levels but we have added more protons to the nucleus and more electrons to the electron cloud. I like to think of it in terms of magnets. Protons and electrons attract each other. The more sets of them you have, the stronger the attraction, and the closer they will be drawn to each other.

Note: The nucleus is massive compared to the size of electrons. Thus, electrons are drawn to the nucleus, not the other way around. A friend of mine once brought up the analogy of a refrigerator and a magnet. A magnet will jump toward a refrigerator, but a refrigerator will stay put.

- **Down a group:** atomic radius gets larger. This trend is easier to understand. The electrons go into a higher energy shell; therefore, they are physically farther from the nucleus.

Ionization energy: the energy cost associated with removing an electron from an atom that is in a gaseous state.

- **Across a period:** ionization energy increases. It gets tougher to remove an electron that is held more tightly.
- **Down a group:** ionization energy decreases. It gets easier to remove an electron that is physically farther from the nucleus. There are also more energy levels filled with electrons between the nucleus and the last valence electron. You might hear the term **shielding** come up. If there are more layers of electrons between the electron to be removed and the nucleus, it's going to be easier to remove that electron because it's better shielded. Ultimately, that means a lower energy cost to removing said electron.

Electron affinity: essentially the opposite of ionization energy. It's the energy change caused by adding an electron to a neutral atom in its gas state. A more negative value means a higher affinity for the added electron.

Note: The trend is generally the same as for ionization energy, but there are always exceptions.

- **Across a period:** electron affinity usually increases. There are more valence electrons as one move across the main groups.

- **Down a group:** electron affinity decreases. As we go down a group, the electrons are physically farther from the nucleus so the attraction for the nucleus is going to be declining.

Valence electrons: those on the outer shell that are available to be shared, gained, or lost on the quest for stability.

Let's step back and think of the big picture.

Metals on the left side of the periodic table have a low number of valence electrons. They would become more stable by losing those electrons. The lower the number of valence electrons, the less likely the atom is going to gain stability by simply gaining an electron. As with many things in chemistry, this is a bit of an oversimplification.

For example: Sodium would become more stable by gaining an electron. It takes energy to lose an electron. However, once the Na^{+1} ion is formed it's attracted to negative ions like Cl^{-1}. That attraction of positive and negative ion makes it stable and forms an ionic bond between the two ions.

Non-metals on the right side of the periodic table have a greater number valence electrons. They become more stable by gaining electrons. Therefore, it's favorable for them to take on an electron and not favorable for them to lose an electron.

Note: Gaining a second electron is not a favorable process apart from the formation of an ionic bond.

Electronegativity: A measure of how strongly a bonded atom attracts electrons to itself. This is an arbitrary scale (a guy named Pauling came up with the scale most used today.) Nevertheless, it's useful for discussing types of bonds formed, which is addressed in a later chapter.

Note: the noble gases are not assigned values because they tend not to bond. They're already stable.

The trends are similar to electron affinity and ionization energy. The closer you get to fluorine, the higher the electronegativity.

- **Across a period:** the electronegativity increases. For example, Fluorine has the highest electronegativity. It's one electron away from being stable, so it's highly motivated to gain one electron. Once it's bonded to another atom, it's quite possessive of the additional electron.
- **Down a group:** the electronegativity decreases. Recall that going down a group the radius is getting bigger, meaning that the last electrons are located farther and farther from the nucleus. This makes it difficult for the nucleus to hang on to its own electrons, let alone attract one in a bonded pair.

Source:
If you want to read more on this, check out https://chem.libretexts.org

Ion Size:

Compared to the neutral atom, positive ions are smaller and negative ions are bigger.

Cations (positive ions):
It's a little easier to understand that positive ions are smaller compared to the neutral atom. Positive ions become who they are by having a neutral atom lose its valence (outer shell) electrons. If Na has three energy levels, and Na^+ has two energy levels, it follows that Na^+ is going to be smaller. You can also think of it in terms of the number of protons and the number of electrons. Before becoming an ion, the neutral atom had an equal number of protons and electrons. After the change to a cation (positive) form, there are fewer electrons than protons.

Another way of thinking about this is to consider the number of energy levels. The electron or electrons lost will come from the valence shell. Therefore, there are less energy levels. Back to our example of Na. Recall the electron configuration for Na is $1s^2\ 2s^2\ 2p^6\ 3s^1$. The highest principle quantum number is three, so there are three energy levels. When Na becomes Na^{+1}, its electron configuration becomes the same as neon: $1s^2\ 2s^2\ 2p^6$. Notice that the highest principle quantum number here is two, so there are two energy levels.

Anions (negative ions):
Atoms turn into negative ions when they gain electrons. Like charges repel, so if you put more electrons in the same general area, they are going to spread out more. You can also think of it in terms of the number of protons vs. the number of electrons. Before the change to the ion, there was an equal number of protons and electrons. After the change to an anion (negative ion), there are more electrons than protons, so there will be an increase in the electrostatic repulsion. Just think of being stuffed into a tight space with your whiny little brother/sister and you will get an idea of how electrons react to other electrons in their personal space.

Caution: Electrons carry a negative charge. When an atom gains electrons, it becomes more negative. This throws a lot of chemistry students off.

Successive Ionization Energy:

When we talk about the trend in ionization energy, we mean the first ionization energy. That is the removal of the first electron from the atom. Successive ionization energy is any one besides the first. So, how much energy does it take to remove a second electron from an atom? What about a third electron? Fourth?

It costs energy to remove an electron from an atom. It costs even more energy to remove a second electron from an atom, and so forth. Removing electrons from the outer energy level (valence shell) is easier than removing electrons from an inner shell. This means that we can count the number of valence electrons by analyzing the successive ionization energies.

An example with completely made up numbers so you get the point:
If it costs 34 kJ/mol of energy to remove the first electron, 54 kJ/mol of energy to remove the second electron, 400 kJ/mol of energy to remove the third electron, and 450 kJ/mol of energy to remove the fourth electron, how many electrons are on the valence shell?

The answer is two. I know this by figuring out the difference between the energy costs to remove each successive electron. Between the removal of the first and second electrons, there's a difference of 20 kJ/mol. Between the second and third there's a difference of 346 kJ/mol, and between the third and the fourth, there's a difference of 50 kJ/mol. You can assume that the largest gap in the amount of energy is going to between the removal of the last valence shell electron and the first inner shell electron.

Chapter 4:
Inorganic Nomenclature (Naming)

Introduction:

This chapter might be getting slowly squeezed out by the great time crunch, but it's still a useful one. If elements are like letters, then compounds are akin to words, and chemical equations are similar to sentences. Let's learn how to write chemical "words" so that later we can fill in the "sentences" of chemical equations.

Definitions:

Ion: what an atom becomes if it loses or gains electrons

Polyatomic ion: molecule with a net charge because it has gained or lost electrons

Common Cations (Positive Ions):

1+		2+	
Ammonium	NH_4^+	Chromium (II)	Cr^{2+}
Copper (I)	Cu^+	Cobalt (II)	Co^{2+}
Gold (I)	Au^+	Copper (II)	Cu^{2+}
Silver	Ag^+	Iron (II)	Fe^{2+}
		Lead (II)	Pb^{2+}
		Mercury (II)	Hg^{2+}
		Nickel (II)	Ni^{2+}
		Tin (II)	Sn^{2+}
		Zinc	Zn^{2+}

3+	4+
Chromium (III) Cr^{3+}	Lead (IV) Pb^{4+}
Cobalt (III) Co^{3+}	Tin (IV) Sn^{4+}
Gold (III) Au^{3+}	
Iron (III) Fe^{3+}	
Nickel (III) Ni^{3+}	

Also these ions from the periodic table:

Lithium Li^{+}	Beryllium Be^{2+}	Aluminum Al^{3+}
Sodium Na^{+}	Magnesium Mg^{2+}	Gallium Ga^{3+}
Potassium K^{+}	Calcium Ca^{2+}	

Common Anions (Negative Ions):

1^{-}	2^{-}	3^{-}
Acetate $C_2H_3O_2^{-1}$	Carbonate CO_3^{2-}	Phosphate PO_4^{3-}
Bicarbonate HCO_3^{-1}	Sulfate SO_4^{2-}	
Hydroxide OH^{-1}	Sulfite SO_3^{2-}	
Nitrate NO_3^{-1}		
Nitrite NO_2^{-1}		
Chlorate ClO_3^{-1}		
Chlorite ClO_2^{-1}		

Also these ions from the periodic table:

Nitride N^{3-}	Oxide O^{2-}	Fluoride F^{-1}
Phosphide P^{3-}	Sulfide S^{2-}	Chloride Cl^{-1}
		Bromide Br^{-1}
		Iodide I^{-1}

Molecular compounds: compounds formed by sharing electrons. The electronegativity difference is small enough that the electrons are considered shared between the two atoms.

Ionic compounds: compounds formed by give-and-take of electrons. The electronegativity difference is great enough (above 1.7) that one atom is considered to have given up its electron(s), so it becomes a positive ion. The other atom it is bonded to accepts the electron(s), becoming a negative ion.

Note: In either case, the positive or negative ion could be polyatomic, meaning it's made up of several atoms that are already sharing electrons and have an overall positive or negative charge.

Binary ionic compounds: ionic compounds that consist of only two types of elements.
Binary molecular compounds: molecular compounds that consist of only two types of elements.

Types of Binary Compounds:

Type 1: a metal and a nonmetal; the metal involved only forms one type of cation (positive ions); Groups 1 and 2 on the periodic table will form positive ions
Type 2: a metal and a nonmetal; the metal involved can form two or more types of cations; transition metals
Type 3: made up of two nonmetals

Note: I wouldn't worry too much about knowing the types of binary compounds. It's just a way to categorize them and break the unit down into smaller, more manageable chunks.

If you want more information, go here:
http://limestone.k12.il.us/teachers/rhebron/Chem_HO/C04_Naming_Writing.html

Steps to Writing Ionic Chemical Formulas:

Step #1: Write down the formulas for the positive and negative ions.
Step #2: Determine the number of each ion needed to get the charges to cancel.
Step #3: Write the number of each ion needed as a subscript to the right of the ion.
Step #4: Add parentheses as necessary.
Step #5: Check your answer.

Annotated Steps to Writing Ionic Chemical Formulas:

Step #1: Write down the formulas for the positive and negative ions. Leave a little space between them. When practicing, you can look these up, but it will be easier if you memorize some common cations (positive ions) and anions (negative ions).

Step #2: Determine the number of each ion needed to get the charges to cancel. Your goal is to get a net charge of zero. That means the total positive charge should be the same magnitude (value) as the total negative charge. They are already opposite in sign.

Step #3: Write the number of each ion needed as a subscript to the right of the ion.

Step #4: Add parentheses as necessary. Parentheses are used when your formula needs more than one of a polyatomic ion.

Step #5: Check your answer.

Criss-cross method of determining the number of ions needed:

Many times it's true that the magnitude of the charge is the same as the number of the other ion you will need to balance it. Just remember to reduce if possible.

Example 1: potassium oxide
Step #1: Write the ions; potassium is K^{+1} and oxide is O^{-2}.
Step #2: To get these two to cancel, I would need two of the potassium ions and one oxide ion.
Step #3: Turn the number of ions into the appropriate subscripts; K_2O.

Note: Do not write in a subscript of one.

Step #4: Parentheses are not needed because neither of these ions are polyatomic.
Step #5: Double check the answer; the overall charge of two potassium ions is +2 and the overall charge of one oxide ion is -2. The net charge is 0, and this is the correct chemical formula for potassium oxide.

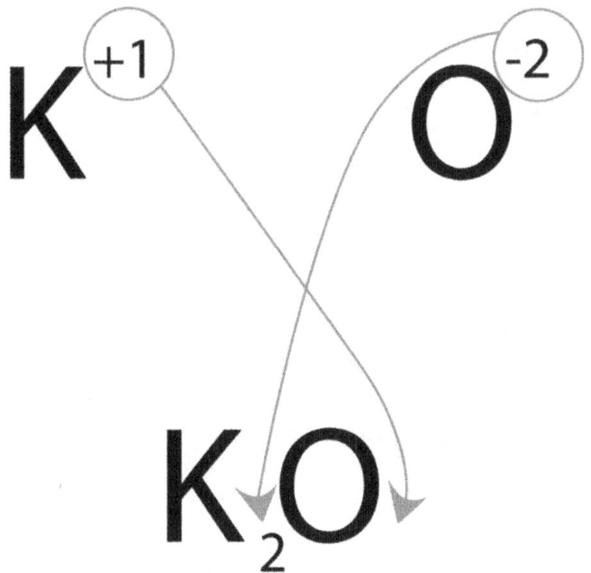

Example 2: gold (III) nitrate
Step #1: Write the ions; gold (III) is Au^{+3} and nitrate is NO_3^{-1}
Note: It is not necessary to write the charge on nitrate as "-1". Having "1-" or just the negative sign is good enough. However, students commonly mistake nitrate as NO with a negative 3 charge when they write only the "-" sign.
Step #2: To get these two to cancel, I would need one of the gold (III) ions and three of the nitrate ions.
Step #3: The formula currently looks like this: $AuNO_{33}$.
Step #4: Parentheses are definitely needed here because it looks like I need 33 oxygen atoms instead of three nitrate ions. The formula should look like this: $Au(NO_3)_3$.

Steps to Naming Ionic Chemical Formulas:

Step #1: Write down the name of the cation (positive ion) and anion (negative ion).
Step #2: Determine if the positive ion can form more than one type of ion. If it cannot, then you are done. If it can, continue to Step 3.
Step #3: Determine the total negative charge for the compound.
Step #4: Determine the total positive charge for the compound.
Step #5: Determine the charge of the positive ion.
Step #6: Check your answer.

Annotated Steps to Naming Ionic Chemical Formulas:

Step #1: Write down the name of the cation (positive ion) first and anion (negative ion) second. Essentially, you're going to have to work backwards. Cations are either metals or polyatomic ions. Anions are either nonmetals or polyatomic ions. Leave a little space in between the two words in case you need to add a Roman numeral later.

Step #2: Determine if the positive ion can form more than one type of ion. If it cannot, then you are done. Adjust for spacing if you left excessive amount of space between the words. If it can, continue to Step 3.

Step #3: Determine the total negative charge for the compound. The negative ion won't change its charge, so you can use that as a starting point. Look at the subscript next to the negative ion and multiply that subscript by the charge of the negative ion to get the total negative charge in this formula.

Note: It is important to distinguish between subscripts that are a part of the chemical formula and subscripts that mean there is more than one of that ion. For example: $Fe_3(PO_4)_2$ has three subscripts. The 3 on the Fe means that there are three of those positive iron (II) ions. The 4 next to the O is a part of the phosphate ion. The 2 outside the parenthesis means that there are two phosphate ions.

Step #4: Determine the total positive charge for the compound. All ionic chemical formulas should be neutral, meaning total positive charge is equal in magnitude to the total negative charge. If the negative charge is -6 then the positive charge must be +6.

Step #5: Determine the charge of the positive ion. Take the total positive charge and divide by the subscript on the positive ion. This will tell you the charge on just one of the positive ions. Write the charge of the positive ion as a Roman numeral to the right of its name.

Step #6: Check your answer.

Example: name $Fe_3(PO_4)_2$
Step #1: Write down the name of the cation (positive ion) and anion (negative ion).

Iron phosphate

Step #2: Determine if the positive ion can form more than one type of ion. If it cannot, then you are done. If it can, continue to Step 3.

Iron can be Fe^{+2} or Fe^{+3}.

Step #3: Determine the total negative charge for the compound. The charge on the negative ion does not change; phosphate will always be PO_4^{-3}, therefore, it will always have a -3 charge. Multiply the charge by any subscript next to the negative ion. There is a subscript of 2 next to the phosphate, so the total negative charge will be -6.

$Fe_3(PO_4)_2$

-3 ← charge of one phosphate ion

-6 ← total negative charge

Step #4: Determine the total positive charge for the compound. If the negative charge is -6 then the positive charge must be +6.

$Fe_3(PO_4)_2$
-3 ← charge of one phosphate ion
Total positive charge →+6 -6 ← total negative charge

Step #5: Determine the charge of the positive ion. Take this total positive charge and divide by the positive ion subscript to get the charge for the positive ion. +6 divided by the subscript of 3 will be +2; therefore, this ion must be Fe^{+2}, iron (II). Write the positive ion charge as a Roman numeral to the right of the positive ion name.

$Fe_3(PO_4)_2$
Charge of one iron ion → +2 -3 ← charge of one phosphate ion
Total positive charge →+6 -6 ← total negative charge

Step #6: Check your answer. The compound name is iron (II) phosphate.

Common Mistakes in Writing and Naming Ionic Chemical Formulas: (no particular order)

- Switching the cation and anion (positive and negative ions). Positive ions or names always come first in ionic chemical formulas.
- Writing CO instead of Co. "C" means carbon and "O" means oxygen. On the other hand, Co is the symbol for cobalt.
- Rampant parentheses. These are used only to indicate more than one polyatomic ion or to contain the Roman numeral when naming a compound with a transition metal that has more than one form.
- Missing parentheses.

- Not reducing. If you have lead (IV) oxide the formula is PbO_2, NOT Pb_2O_4.
- Needlessly reducing. Poor sulfate has had its number of oxygen slashed more times than I can count. Leave sulfate alone, young chemistry meanies! It's no longer sulfate if you change its chemical formula. Gold (I) sulfate is written Au_2SO_4 NOT $AuSO_2$.
- Missing subscripts. Subscripts are your friends; they bring balance to the poor formulas-to-be.
- Mistaking the subscript in a polyatomic ion for the subscript denoting the number of ions. Sulfate ceases to be sulfate if you change the number of oxygen in its formula. Sulfate = SO_4^{2-}. Sulfite = SO_3^{2-}.

Naming Binary Molecular Compounds:

Steps to naming binary molecular compounds:

Step #1: Learn the prefixes and the # of atoms they represent

Prefix	# of atoms	prefix	# of atoms
mono	1	hexa	6
di	2	hepta	7
tri	3	octa	8
tetra	4	nona	9
penta	5	deca	10

Step #2: The first element (less electronegative) goes by its name, preceded by a prefix starting at di. If there is just one of them, there is no prefix.

Step #3: The second element is preceded by a prefix and has its ending changed to -ide.

Note: Adjust for special cases. If there are any cases where two "a's" or two "o's" or "a" and "o" would be right next to each other, drop the "a" or "o" from the prefix. In general, try to keep from putting two like vowels together, except "i."

Examples:
1. Cl_2O = dichlorine monoxide
2. NI_3 = nitrogen triiodide
3. N_2H_4 = dinitrogen tetrahydride

Practice Problems:
1. BrF_5 =
2. I_2O_5 =
3. OF_2 =
4. PBr_3 =
5. P_4S_3 =
6. As_4O_6 =
7. As_2O_5 =
8. $AsCl_3$ =
9. ClF_3 =
10. ClF_5 =
11. Cl_2O_7 =

Answers to Practice Problems:
1. BrF_5 = bromine pentafluoride
2. I_2O_5 = diiodine pentoxide
3. OF_2 = oxygen difluoride
4. PBr_3 = phosphorus tribromide
5. P_4S_3 = tetraphosphorus trisulfide
6. As_4O_6 = tetrarsenic hexoxide
7. As_2O_5 = diarsenic pentoxide
8. $AsCl_3$ = arsenic trichloride
9. ClF_3 = chlorine trifluoride
10. ClF_5 = chlorine pentafluoride
11. Cl_2O_7 = dichlorine heptoxide

Practice going from name to formula too.

Practice quiz naming binary molecular formulas:
www.pafaculty.net/biology/keith/kr_graph_site/molec_nomenclature_page.htm

Conclusion:

Learning to write chemical formulas and recognize their names will help in future discussions about chemical reactions. We're in the midst of the chemistry chapters that flow right into one another. From here, we'll head into chemical reactions then on to mole concepts and stoichiometry. Stoichiometry will then come back in gases and solutions. If you're shaky on these foundational chapters, take the time to review them. It's usually just a matter of practice.

Chapter 5: Chemical Bonding and Molecular Structure

Introduction:

After spending some time getting familiar with the different naming systems, it makes sense to look at the types of bonds in more detail. Some of the models will let us gain a better picture of the shapes of certain molecules.

Definitions:

Compound: a substance made up of two or more elements

Does a compound have the same characteristics as the atoms that made it up?

No. For example, Sodium (Na) has one valence electron, which means it's fairly highly motivated to lose one electron. Chlorine (Cl) has seven valence electrons, which means it's fairly highly motivated to gain one electron. So, atomic Na and Cl are very reactive. However, once the transaction has taken place and sodium has given chlorine its valence electron, both are ions with stable octets. This makes them unlikely to react further because a strong bond exists between them thanks to the attraction of the opposite charges. Incidentally, this is also how an ionic bond is formed.

Atoms with extra valence electrons will donate them to atoms that need extra electrons. The ratio of how many atoms will donate and how many will gain electrons depends on the overall charge of the ions formed after the transfer(s) have taken place.

Bond: an attraction between two atoms, ions, or molecules

Electronegativity: an atom's tendency to attract a bonding pair of electrons. Fluorine has the highest electronegativity. The trend is that up a group and across a period, the electronegativity gets greater. Noble gases are not assigned electronegativity values because they do not readily form bonds.

Ionic bond: formed by the transfer of electron(s) between atoms, forming ions. The atom that gave up one or more electrons will become positively charged. The atom that accepts one or more electrons will become negatively charged. Think of it as one atom stealing electron(s) from another.

Covalent bonds: a bond formed by sharing of electrons

How does electronegativity difference determine the type of bond formed?

When speaking of type of bond in this section, I'm referring to non-polar covalent, polar-covalent, and ionic. These are based off of the difference in electronegativity. Depending on which source you look at, the cutoff points could be different. An electronegativity below .4 could be considered non-polar covalent. An electronegativity difference between .4 and 1.7 could be considered polar-covalent. An electronegativity difference above 1.7 could be great enough to be considered ionic.

You will have to ask your teacher what to count electronegativity differences that fall on the cutoff lines (.4 and 1.7). I usually tell my students to consider it the higher type of bond, polar covalent and ionic respectively.

Think of it like a tug of war for electrons.

Example 1: If two elements are the same, like F_2, the electronegativity difference would be 0. This would form a non-polar covalent bond.

Translation: Two fluorine atoms will share a pair of electrons very evenly, so there will not be a pole.

Note: "Pole" refers to an area of positive charge or an area of negative charge.

Example 2: If hydrogen (H), electronegativity of 2.1 and oxygen (O), electronegativity of 3.5 formed a bond, what type of bond would it be? 3.5 – 2.1 = 1.4 An electronegativity difference of 1.4 is great enough to be considered polar covalent, but slightly smaller than what would be considered ionic.

Example 3: If sodium (Na), electronegativity of 0.9 and chlorine (Cl), electronegativity of 3.0 formed a bond, what type of bond would it be? 3.0-0.9 = 2.1. An electronegativity difference of 2.1 is great enough to be considered ionic.

Shared pair: a pair of electrons that can count for the valence shell of both atoms it lies between; these are represented by drawing a line between the atoms that are sharing them

Lone pair: a pair of electrons that belongs to the valence shell of only one atom; these are represented by drawing two electrons next to the symbol for the atom to which they are assigned

Note: When referring to "type" of bond, we could also just be asking a question about how many shared pairs of electrons lie between the bonded atoms.

How does the number of shared pairs of electrons relate to the bond strength?

Single bond: a shared pair of electrons. These are the longest and weakest type of bond.
Double bond: two shared pairs of electrons between the same two atoms. These are shorter than single bonds and longer than triple bonds. They are also stronger than single bonds and weaker than triple bonds.
Triple bond: three shared pairs of electrons between the same two atoms. These are the shortest and strongest of the types of bonds that refer to number of shared pairs of electrons.

Note: By "strongest" I mean it will take the most energy to break this type of bond.

Note: There is no quadruple bond.

Review of Steps to Drawing Lewis-dot (electron-dot) Structures for Elements:

1. Look at a Periodic Table of the elements and find which group the element in question is located. Most of the ones you'll be given will be from the main groups (1, 2, 13, 14, 15, 16, 17, 18).
2. Determine how many valence electrons the element should have.

Tip: The number of valence electrons = the last digit of the main group number. For example, Li is in group 1 it has 1 valence electron. S is in group 16, it has 6 valence electrons.

3. Draw the symbol for the element and draw one dot to represent each valence electron.

Why would two atoms need to share more than one pair of electrons?

When we get to drawing Lewis-dot structures shortly, we will need to create substances that meet two criteria.

1: account for the correct number of valence electrons
2: obey the octet rule where applicable (there are some exceptions to the octet rule)

The first example I go through with my students to illustrate the importance of the two rules and introduce single, double, and triple bonds is to have them draw Lewis-dot structures for molecular fluorine, oxygen, and nitrogen.

Example 1: fluorine, F_2 (it is diatomic)
Each fluorine atom has seven valence electrons, so there are fourteen electrons that need to be accounted for. Both fluorine atoms are one electron shy of a stable octet. However, if they share an electron pair between them, forming a non-polar covalent bond, they can count both electrons for their valence shell. If you fill in the rest of the lone pair of electrons, and count, you will come up with fourteen, which is good.

Note: Every lone pair counts for two electrons. Every shared pair of electrons counts for two electrons.

Example 2: oxygen, O_2 (also diatomic)
Each oxygen atom has six valence electrons, so the Lewis-dot structure that is formed should account for twelve valence electrons. Both oxygen atoms need two more valence electrons to get a stable octet. If we only put one shared pair of electrons between them, we would need fourteen electrons in total to fill out the octets. There are not fourteen electrons available, only twelve. The best way to solve that problem is to share two pairs of electrons, forming a double bond. If you share two pairs then fill out the octets, you will end up using twelve electrons exactly.

Example 3: nitrogen, N_2 (you guessed it, diatomic) Each nitrogen atom has five valence electrons, so the Lewis-dot structure that is formed should account for ten valence electrons. Both nitrogen atoms need three more valence electrons to fill out their octets. If we only put one shared pair of electrons between the nitrogen atoms, we would need fourteen electrons in total to fill out the octets. We don't have fourteen electrons. We have ten. Recall that sharing two pairs of electrons allowed us to use twelve electrons. It follows then that if we share three pairs of electrons between the nitrogen atoms, we will need ten electrons to complete our structure.

Steps to Drawing Lewis-dot Structures for Basic Molecular Compounds:

Step #1: Count the valence electrons needed for the structure.
Step #2: Write down the central atom and connect it once to every other atom.
Note: The central atom is the least electronegative one (the one written first) with the exception of hydrogen.

Why can't hydrogen be the central atom?
It has only one valence electron, so it can only form one bond. The central atom must be capable of forming more than one bond.

Step #3: Add lone pairs of electrons to every atom that needs a complete octet. The most common exceptions to the octet rule are hydrogen (H), beryllium (Be), and boron (B).

Note: Hydrogen never gets lone pairs of electrons. Ever.

Step #4: Check to see if your structure is accounting for the correct number of electrons.

Example 1: Draw the Lewis-dot structure for H_2O
Step #1: Count the valence electrons. H is in group 1, so they will both have 1 valence electron. O is located in group 16, so it will have 6 valence electrons. (1 + 1 + 6 = 8 valence e^-)

Step #2: Put the central atom down and connect each other atom with one bond. Oxygen (O) is the central atom by default since H can't be central. Bond it once to each hydrogen (H) atom.

Step #3: Fill the octets as necessary. H does not need a stable octet. O will need two lone pairs to complete its octet because it already has a bond (shared pair of electrons) to each hydrogen.

Step #4: Check to see that your structure has used the correct number of electrons. 2 shared pairs, 2 lone pairs = 8 electrons.

Steps to Drawing Lewis-dot Structures for Polyatomic Ions:

I will use sulfite (SO_3^{-2}) as an example.
Step #1: Count the valence electrons. The electrons that must be accounted for are slightly different because you will need to add or subtract the charge. People get tripped up here because a charge of +2 means that two electrons have been lost and a charge of -2 means two electrons have been added. Sulfite's valence electrons: (6 from S) + (3 x 6 from O) + (2 from charge) = 26 ve^-.
Step #2: Write the central atom and bond each other atom to it once. S is central and it will be bonded once to each O.
Step #3: Fill the octets. Each O needs three lone pairs of electrons. S needs one lone pair since it is only bonded three times, once to each O.
Step #4: Count the electrons used. If you count the eight electrons around each of the three O atoms, including the shared bonds with S, (8 x 3 = 24 e-) and add them to the two electrons from the lone pair on S, you get a total of twenty-six.

Note: This structure is correct but does not consider the topics of resonance or formal charge. (My main concern is that the correct # of electrons is used and the octets are filled. In layman's terms, resonance means that there is more than one legitimate structure and the "real" structure is actually a hybrid of those.)

If you would like an additional worksheet concerning drawing Lewis-dot structures, there are hundreds on the internet. You can also email me a request at juliecgilbert5Steps@gmail.com. Please include the word "Chemistry" in the subject line so I can distinguish it from other 5 Steps books.

The VSEPR Theory and Molecular Shape:

VSEPR stands for **V**alence **S**hell **E**lectron **P**air **R**epulsion. The definition is pretty much in the acronym. Think of it as electrons really don't like each other and stay as far away from each other as is physically possible.

- If you have two atoms attached to a central atom, the biggest bond angle you could get is 180°. This would be a linear structure.
- If you have three atoms attached to a central atom, the biggest bond angle you could get is 120°. This would be a trigonal planar structure.
- If you have four atoms attached to a central atom, you have to move to three dimensional space to get the biggest bond angle possible. If you form a tetrahedral shape, you can get bond angles of 109.5°. Otherwise, you would have a maximum of 90° angles.
- If you have five atoms attached to a central atom, you would form a trigonal bipyramid in order to keep them as far from each other as possible. A trigonal bipyramid is like having two pyramids lined up back to back with one facing up and another down. The corners where they meet would be a trigonal planar shape. Then there would be a linear structure running straight through it.

- If you have six atoms attached to a central atom, you would form an octahedral shape. It's like a square planar structure with a linear structure stabbing through the center of it.

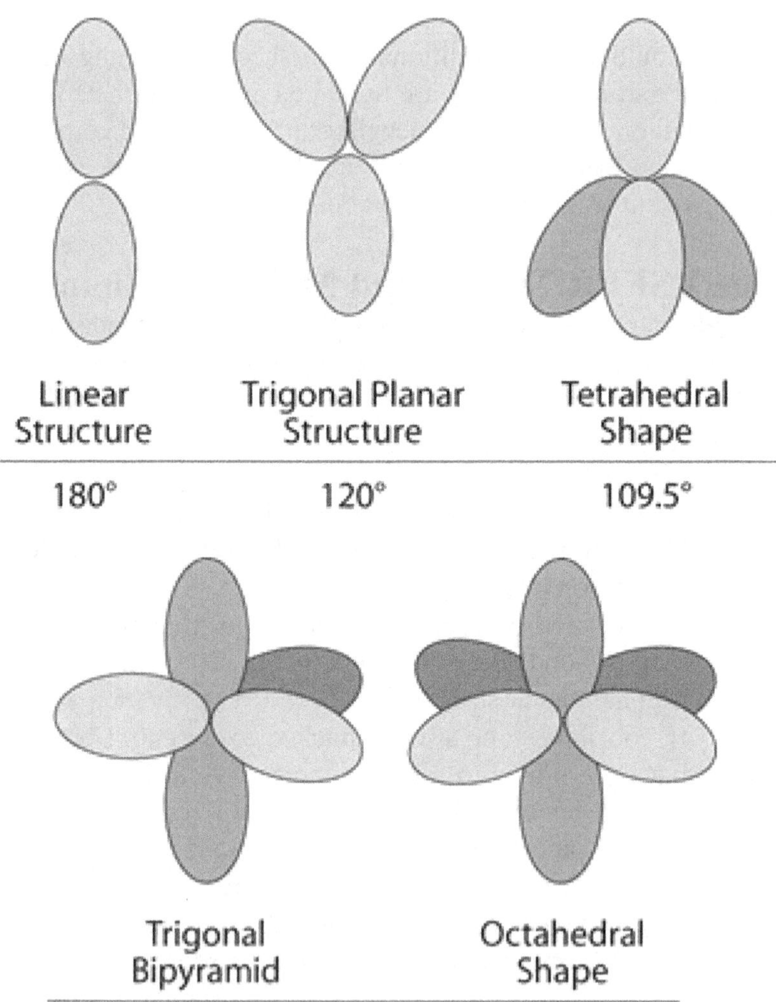

There's a difference between the electron domain geometry and the molecule shape. Electron domain geometry refers only to places where there are electrons, be they shared pairs or lone pairs. However, when considering the molecule's shape, we see only the effect of the lone pairs of electrons, not the electrons themselves.

For example, a water molecule (H_2O) has four electron domains, but of those, only two are shared pairs of electrons running between O and H. The electron domain geometry is tetrahedral, but the actual molecule shape is bent.

Note: You can get a bent structure with trigonal planar electron domain geometry if one of the locations is a lone pair. The difference between the resulting bent structure from trigonal planar and tetrahedral electron domain geometry is the bond angle of the final molecule (~120° and ~109.5° respectively).

Molecular Polarity:

In order for a molecule to be polar, it needs "poles." In other words, it needs to have an area that is partially positive and an area that is partially negative. For this to happen, you need to have polar covalent bonds and molecular geometry that supports polarity.

For example, CO_2, has polar bonds because carbon (C) has an electronegativity of 2.5 and oxygen (O) has an electronegativity of 3.5, but the molecular shape is linear.

Without dots, the structure looks like this: O = C = O

With dots, the structure looks like this: :Ö=C=Ö:

The oxygen (O) atoms are more electronegative, so the O atoms will pull the electrons toward themselves. If the left oxygen pulls left and the right oxygen pulls right, their effect upon the electrons will cancel out. There's no place I could cut the molecule in half and say, this half is negative and this half is positive. Therefore, there are no poles and the molecule is non-polar.

Intermolecular forces:

Intramolecular forces: forces of attraction that hold a molecule or compound together; "intra" means inside
Intermolecular forces: temporary attractions that exist between molecules; "inter" means between

Types of Intermolecular Forces:
London dispersion forces (LDF): momentary, induced dipoles formed by a temporary shift in electrons; all molecules will have this type of intermolecular force.
Dipole-dipole forces: this type of attraction lies between molecules that have permanent dipoles. "di" means two, so this refers to a polar molecule like HCl. The electronegativity of H is 2.1, and the electronegativity of Cl is 3.0. The difference between those is 0.9, which is considered strong enough to be polar. The electrons will dwell closer to the Cl, so it picks up a partial negative charge. The electrons are pulled away from the H, so it picks up a partial positive charge. Because I could draw an imaginary line between the H and the Cl, separating the positive half from the negative half, I know HCl is a polar molecule.

Ion-dipole forces: this type of attraction lies between polar molecules and ions. It's stronger than a dipole-dipole force because ions are fully charged, therefore they have stronger pull.

Hydrogen bonds (H-bonds): a particularly strong type of dipole attraction formed between a hydrogen and something that's highly electronegative like fluorine (F), oxygen (O), or nitrogen (N).

How does bonding affect the behavior of a compound?

Network solid: atoms are connected by a series of continuous covalent bonds throughout the structure. The numerous connections between the atoms makes these molecules pretty hard.

Metallic bonding: a force of attraction that exists between valence electrons and metal atoms. Many metal atoms will share electrons. The valence electrons are said to be delocalized, meaning they don't belong to any one particular metal atom. This free-flowing sea of electrons is what gives metals some of their properties like the ability to conduct electricity.

Ionic solid: formed from ions arranged in a crystal structure; these solids are hard but brittle

Visit this site if you want a more thorough look at the types of solids:
https://www.chem.fsu.edu/chemlab/chm1046course/solids.html

Chapter 6:
Chemical Reactions and Balancing Equations

Introduction:

Balancing chemical equations is a skill you'll need moving forward. Many times in future units you will be given the balanced chemical equation, but not always. Sometimes, you will be given the number of grams of two things reacting together and asked to predict the amount of product that could be formed. Such a calculation cannot be completed without first balancing the equation. Besides, it can be kind of fun.

Anatomy of a Balanced Chemical Equation:

$Pb(NO_3)_{2(aq)} + Na_2O_{(aq)} \rightarrow 2\ NaNO_{3(aq)} + PbO_{(s)}$

Arrow: yields or produces; located in between the reactants and products
Reactant/reactants: located left of the arrow; substances participating in the chemical reaction; $Pb(NO_3)_{2(aq)} + Na_2O_{(aq)}$
Product/products: located right of the arrow; substances resulting from a chemical reaction; $NaNO_{3(aq)} + PbO_{(s)}$
Subscript: Subscripts can mean different things. Some are part of the formula for a polyatomic ion and some are telling you the number of atoms or ions. For example, $Pb(NO_3)_{2(aq)}$ has a

subscript of 3 on the oxygen and a subscript of 2 outside the parentheses. The 3 in NO_3 means there are three oxygen atoms as a part of the polyatomic nitrate ion (NO_3). The subscript of 2 outside the parenthesis says that in this particular chemical formula, $Pb(NO_3)_2$, there are two nitrate ions.

Chemical formula: describes how many atoms/ions are in a substance

Coefficient: tells how many molecules or formula units of a substance are needed

Note: You never write in a coefficient of 1.

State of matter: symbol showing what state the substance currently exists in

Helpful hints to balancing equations:

1. Balance the different types of atoms one by one
2. First balance atoms that only show up once on each side
3. Balance polyatomic ions that appear on both sides as a single unit (you may use a substitution method if it is easier for you)
4. Balance H and O after all other atoms. These typically show up the most, thus are most difficult to balance. As a general rule of thumb, if an element is in more than one place on a given side of the equation, save it for last.

Key point: "balanced" means there are the same number of each element on each side

There are two methods for balancing chemical equations: inspection and inventory.

Inventory:

Step #1: Set up an inventory. Draw a line down from the arrow to separate the reactants and products. Write the symbol for each element (or polyatomic ion) present on both sides of the equation. Reactants go left of the line you just drew and products go right of the line you just drew. You don't have to line them up, but I think it makes your life easier to do so. Note the subscripts next to each element or polyatomic ion and write that number next to the element symbols.

Step #2: Pick one element (or polyatomic ion) and determine what you need to get them to balance. Remember, "balance" means there are the same number of each element on each side. The number you determine you need to multiply a certain compound by gets written to the left of the compounds as a coefficient.

Tip: If a polyatomic ion is on both sides of the equation, leave them together. It makes those elements easier to track.

Step #3: Be sure to adjust your inventory to reflect the coefficient's effect on the entire compound.

Step #4: Repeat steps #2 and #3 until each element or polyatomic ion is balanced.

Example 1: ___ $Sb_{(s)}$ + ___ $O_{2(g)} \rightarrow$ ___ $Sb_2O_{3\ (s)}$
Step #1: Inventory
___ $Sb_{(s)}$ + ___ $O_{2(g)} \rightarrow$ ___ $Sb_2O_{3\ (s)}$

1 Sb	2 Sb
2 O	3 O

Step #2: Balance one atom or ion
If I multiply the left by three and the right by two, I can get both sides to have six oxygen atoms.
___ $Sb_{(s)}$ + _3_ $O_{2(g)} \rightarrow$ _2_ $Sb_2O_{3\ (s)}$

1 Sb	2 Sb
6 ~~2~~ O	~~3~~ O 6

Step #3: Adjust the inventory to reflect the coefficient's effect on the entire compound.
___ $Sb_{(s)}$ + _3_ $O_{2(g)} \rightarrow$ _2_ $Sb_2O_{3\ (s)}$

1 Sb	~~2~~ Sb 4
6 ~~2~~ O	~~3~~ O 6

Step #4: Balance the rest of the equation.
$\underline{\;4\;}\ Sb_{(s)} + \underline{\;3\;}\ O_{2(g)} \rightarrow \underline{\;2\;}\ Sb_2O_{3\,(s)}$

4 ~~1~~ Sb	~~2~~ Sb 4
6 ~~2~~ O	~~3~~ O 6

Final answer: $\underline{\;4\;}\ Sb_{(s)} + \underline{\;3\;}\ O_{2(g)} \rightarrow \underline{\;2\;}\ Sb_2O_{3\,(s)}$

Note: You don't need to put a line, but you do need enough space for a coefficient. On a test or a quiz, you may see a line because your teacher doesn't want to tell you which elements/compounds get a coefficient and which do not.

Inspection is pretty much the same thing except that you don't keep an inventory list below the equation. It's more of a mental analysis. I think they're kind of fun. Look at the helpful hints, choose an element (or polyatomic ion) and let it change other things as you go along.

Example 2:
$\underline{\;2\;}\ AgI + \underline{\quad}\ Fe_2(CO_3)_3 \rightarrow \underline{\quad}\ FeI_3 + \underline{\quad}\ Ag_2CO_3$
There's no right or wrong way to start balancing by inspection, but there are sometimes "better" or "easier" ways. Let's start with silver. There is one Ag on the left and two Ag on the right. I would put a 2 next to the AgI. This changes the amounts of I on the left to two.

$\underline{\;6\;}\ AgI + \underline{\quad}\ Fe_2(CO_3)_3 \rightarrow \underline{\;2\;}\ FeI_3 + \underline{\quad}\ Ag_2CO_3$

There are three I's on the right. To fix this, I'd need to multiply the left by 3 and the right by 2 to make them both 6. Since I already have a 2 in front of AgI, I'm going to change it to a 6 because 2 x 3 = 6. At the same time, I'm going to put a 2 in front of FeI_3. The I's should be balanced at this point.

Putting a 2 in front of FeI_3 gave me two Fe on the right side of the equation. This is good because there are two Fe's in

$Fe_2(CO_3)_3$ on the left.

_6_ AgI + ___ $Fe_2(CO_3)_3$ → _2_ FeI_3 + _3_ Ag_2CO_3
Consider the CO_3's. There are three on the left, so there should be three on the right as well. To get three on the right, I need to put a 3 in front of the Ag_2CO_3. Incidentally, this also fixes the amount of Ag on the right.

Final answer: 6 AgI + $Fe_2(CO_3)_3$ → 2 FeI_3 + 3 Ag_2CO_3

Five Types of Chemical Reactions:

Synthesis: two substances coming together to form one product
Decomposition: one substance breaking down into two or more products
Single replacement (single displacement): an element and a compound reacting to form a different element and a different compound
Double replacement (double displacement): two compounds reacting to form two different compounds
Combustion: a compound reacting with oxygen that results in the release of energy
Note: in my college prep course, we stick to general hydrocarbons so the resulting products are carbon dioxide and water

Five types of chemical reactions in letter form:
Synthesis: A + B → AB
Decomposition: AB → A + B
Single replacement (single displacement): A + BC → B + AC or X + YZ → Z + YX
Note: the difference between these two forms is whether the positive ion is being replaced or the negative ion is being replaced.
Double replacement (double displacement): AB + CD → AD + CB
Note: C must come before B on the products side because it is a positive ion

Combustion: $C_xH_y + O_2 \rightarrow CO_2 + H_2O$
Note: this is not the only general form for a combustion reaction.

Identifying the Five Types of Chemical Reactions:

I call these the "marks" of each type of reaction. They're the things you should look for if you're asked to identify the type of chemical reaction.
Synthesis: one product
Decomposition: one reactant
Single replacement (single displacement): one element, one compound
Double replacement (double displacement): two compounds
Combustion: usually a hydrocarbon (a compound composed solely of hydrogen and carbons); definitely oxygen (O_2)

Special Topics in the Chemical Reactions Chapter:
You will likely have to learn to read solubility charts and activity series.
Solubility charts: The most common types of these will have the positive ions written on the left side and the negative ions written across the top. It might help to place a finger on the positive ion on the left and the negative ion on the right. Next, trace right and down until your fingers meet. If it says "s" in the box, it means that substance is not soluble in water. If there is "aq" written in the box, it means that the substance is water soluble.

These are used in double replacement reactions.

Activity series: a list of metals (and hydrogen) according to their reactivity. Many will have the more reactive elements written on top and then descending toward less reactive elements. There are activity series for nonmetals, but they're much smaller. It's generally the same as group 17 written top to bottom. Fluorine is more reactive than chlorine, which is more reactive than bromine, which is more reactive than iodine.

These are used in single replacement reactions.

In order for the single replacement reaction to take place the element (the thing alone by itself) must be higher on the activity series in order for the reaction to happen.

Note: Sometimes an activity series is written horizontally. If that's the case, there will be an arrow telling you which are more reactive.

Common mistakes balancing chemical equations: (no particular order)

General:

- Not correctly identifying the type of reaction; combustion often mistaken for single replacement
- Mistaking the roman numerals for an indicator of the subscript
  For example, Iron (III) does **NOT** translate to Fe_3 in a chemical equation
- Confusing ions; nitrate (NO_3^{-1}) and nitride (N^{-3}) are commonly interchanged
- Not naming the ionic compounds correctly; for example, in lead (II) chloride don't forget the roman numerals

Synthesis and Decomposition:

- Forgetting which elements are diatomic (N_2, O_2, F_2, Cl_2, Br_2, I_2, H_2)
- Forgetting when elements are diatomic. Elements are only diatomic when they are in their free element state, meaning gas.
- Forgetting when elements are becoming ions and must obey ion charge rules.
  Example 1: Synthesis of aluminum + chlorine →
  $4\ Al + 3\ Cl_2 \rightarrow 2\ AlCl_3$
  Free elements → ionic compound
  Example 2: Decomposition of sodium chloride →

$2\ NaCl \rightarrow 2\ Na + Cl_2$
Ionic compound → free elements

Single Replacement and Double Replacement:

- Replacing the wrong element/ ion; two positive ions will NEVER form a chemical compound
- Not balancing the individual chemical formulas
- Forgetting to put the state of matter where applicable

Combustion:

- Forgetting to balance water's oxygen first
  Note: This doesn't always apply with alcohols and other organic compounds that have additional oxygen as a part of their chemical formula

Chapter 7: Mole Concepts

Introduction:

This chapter contains a lot of different topics, but they all revolve around the mole. It's probably going to be a long chapter, but skim the titles of each section if you're just looking for something specific.

What is a mole?

The mole is a very important unit. It's been called the chemist's dozen. At this point in your academic career, you should be familiar with the concept of a dozen. A dozen is twelve items. A mole is 6.022×10^{23} items. In chemistry, those items are called atoms, molecules, ions, or simply particles if you want to generalize more.

Moles allow for conversions between the number of particles and the mass of those particles. Some students have a hard time grasping numbers as large as 6.022×10^{23}. So, let's return to the concept of a dozen to grasp the relationship to mass. If I have a dozen (twelve) identical pencils, they will have a certain mass. You could measure the mass on a balance. If I have a dozen (twelve) identical cars, they too will have a certain mass. If we had a large enough scale, we could measure the mass of the cars. The number of the items in both cases is a dozen (twelve), but the masses will be vastly different. The same holds true for atoms, ions, and other particles. A hydrogen atom has a mass of 1.00 g. A gold atom has a mass of 197.0 g. So, a mole of hydrogen atoms and a mole of gold atoms would both contain 6.022×10^{23} number of atoms, but the masses of those would be vastly different.

Definitions:

Formula unit vs. molecule:
Formula unit: the simplest ratio of ions in an ionic compound
Molecule: made of two or more atoms that are covalently bound to one another
Note: same idea, different term based on the type of compound you're dealing with

Atomic mass units (amu) vs. grams (g):
For all intents and purposes, they are equivalent units of mass.
Note: you might see it referred to as u for unified atomic mass unit or Da for Dalton. There's a long history of naming units after famous dead guys (notable scientists).

Atomic mass units: one atomic mass unit is 1/12 of the mass of a Carbon-12 atom
Note: There is a technical difference in the mass of amu and grams, but in the amount of significant figures we deal with on the periodic table the difference is negligible (too small to impact the results).

Formula mass vs. Molar mass:
Formula mass: the mass in one formula unit of a substance
Molar mass: the mass per mole of a substance

The average atomic mass found on the periodic table is usually accepted as the mass per mole of that element.

Steps to calculating formula mass and molar mass:

Step #1: Write the balanced chemical formula for the compound
Step #2: Write down the number of each type of atom
Step #3: Convert from the number of atoms to the mass by using the average atomic mass found on the periodic table of elements.
Step #4: Add the values together and report the final answer.
Note: There's a long and a short form of this concept. I'll show you the long form once so you can see what's going on, but then I usually don't ask to see long form.
Note: Molar mass is obtained the same way.

Example: What is the formula mass for calcium chloride?
Step #1: $CaCl_2$
Step #2: 1 atom of Ca and 2 atoms of Cl
Step #3:
1 Ca atom x $\frac{40.1\ amu}{1\ Ca\ atom}$ = 40.1 amu
2 Cl atoms x $\frac{35.5\ amu}{1\ atom\ Cl}$ = 71.0 amu
Step #4: 40.1 amu + 71.0 amu = 111.1 amu

Example: What is the formula mass for calcium chloride? (Short form)
Step #1: $CaCl_2$
Step #2: 1 atom of Ca and 2 atoms of Cl
Step #3:
1 Ca atom x 40.1 amu = 40.1 amu
2 Cl atoms x 35.5 amu = 71.0 amu
Step #4: 40.1 amu + 71.0 amu = 111.1 amu

Quick review of conversions:
Step #1: Start with given information
Step #2: Put units for the unit that needs to change on the bottom and units that you're changing to on the top
Step #3: Repeat step 2 until you get to the final unit
Step #4: Plug in the conversion factors
Step #5: Solve (top = multiply, bottom = divide)

Steps to using molar mass as a conversion factor:

Step #1: Write down the number, unit, and label for the starting compound.
Step #2: Put the unit that matches the starting compound on the bottom of the conversion.
Step #3: Put the unit that you'd like to change to on the top of the conversion.
Step #4: Repeat steps 3 and 4 until you've changed completely over to the desired unit and label
Step #5: Solve and check your answer

Note 1: It's the same steps for formula mass being used as conversions, just with different units.
Note 2: By "label" I mean the chemical formula.
Note 3: For obvious reasons, I don't let my students completely abbreviate formula units. (It's just not polite.) I make them write form u if they want to abbreviate it.

Example 1: a one-step problem
How many moles are equal to 2.55 x10^{28} formula units of lithium carbonate?
Step #1: 2.55 x 10^{28} formula units Li_2CO_3
Step #2: Put the unit that matches the starting compound on the bottom of the conversion.

$$2.55 \text{ x}10^{28} \text{ form units } Li_2CO_3 \text{ x } \frac{}{6.02 \; x \; 10^{23} form \; u \; Li2CO3}$$

Step #3: Put the unit that you'd like to change to on the top of the conversion.

$$2.55 \text{ x}10^{28} \text{ form units } Li_2CO_3 \text{ x } \frac{1 \; mol \; Li2CO3}{6.02 \; x \; 10^{23} form \; u \; Li2CO3} =$$

Step #4: Repeat steps 3 and 4 until you've changed completely over to the desired unit and label. Not necessary here since I'm already in the unit I want to be in after this first conversion.

Step #5: Solve and check your answer.

2.55×10^{28} form units Li_2CO_3 x $\frac{1\ mol\ Li2CO3}{6.02\ x\ 10^{23} form\ u\ Li2CO3}$ =

42358.8039867 mol Li_2CO_3

To three sig figs, the answer would be 42400 mol Li_2CO_3 or 4.24 x 10^4 mol Li_2CO_3.

Example 2: a two-step problem

How many g are equal to $2.55\ x10^{28}$ formula units of lithium carbonate?

Step #1: $2.55\ x10^{28}$ formula units Li_2CO_3

Step #2: Calculate the molar mass of the compound and set up a conversion.

The molar mass of Li_2CO_3 = 73.8 g/mol

Step #3: Put the unit that matches the starting compound on the bottom of the conversion.

$2.55\ x10^{28}$ form units Li_2CO_3 x $\frac{}{6.02\ x\ 10^{23} form\ u\ Li2CO3}$

Step #4: Put the unit that you'd like to change to on the top of the conversion.

$2.55\ x10^{28}$ form units Li_2CO_3 x $\frac{1\ mol\ Li2CO3}{6.02\ x\ 10^{23} form\ u\ Li2CO3}$

Step #5: Repeat these steps until you've changed completely over to the desired unit and label

$2.55\ x10^{28}$ form units Li_2CO_3 x $\frac{1\ mol\ Li2CO3}{6.02\ x\ 10^{23} form\ u\ Li2CO3}$ X $\frac{73.8\ g\ Li2CO3}{1\ mol\ Li2CO3}$

Step #6: Check your answer.

2.55 x 10^{28} form units Li_2CO_3 x $\frac{1\ mol\ Li2CO3}{6.02\ x\ 10^{23} form\ u\ Li2CO3}$ X $\frac{73.8\ g\ Li2CO3}{1\ mol\ Li2CO3}$ = 3.13×10^6 g Li_2CO_3

Reminder: Compounds are created from elements combined in ratios of small whole numbers.

Tip: The following is something I have all my students write on a notecard. If it helps, great. If it confuses you, ignore it. This is a summary of the types of conversions you will need to do.

$\frac{1\ mole}{g}$ $\frac{6.022\ x\ 10^{23} particles}{1\ mole}$

---------------> ------------------->

Mass ← molar mass → Moles ← Avogadro's Number → Particles

<--------------- <------------------

$\frac{g}{1\ mole}$ $\frac{1\ mole}{6.022\ x\ 10^{23} particles}$

Explanation:
You could be asked to convert from mass to moles, moles to particles, or mass to particles, or be asked to go the other way from particles to moles, moles to mass, or particles to mass. The card should help you figure out which conversion to use and what it looks like.

Steps to solving percent composition (% by mass) problems:

Step #1: Write down the chemical formula for the compound. (Sometimes, you may be given this information.)
Step #2: Find the mass of each element in the compound.
Step #3: Add the answers to get the total mass of the compound.
Step #4: For each element, take the mass of that element and divide it by the total mass of the compound.
Step #5: Multiply by 100 to turn this number into a percent.
Step #6: Check your answer by adding the percent values together. They should be very close to 100%.

Example: What is the percent composition of aluminum sulfate?
Step #1: chemical formula: $Al_2(SO_4)_3$
Step #2: Find the mass of each element in the compound.
2 Al x 27 g = 54 g
3 S x 32.1 g = 96.3 g
12 O x 16.0 g = 192 g

Tip: Leave extra room next to oxygen because; it's tempting for some students to think it's a zero.

Note: Your periodic table may have different values; I'm using one that only has one decimal place.

Step #3: Add the answers to get the total mass of the compound.
2 Al x 27 g = 54 g
3 S x 32.1 g = 96.3 g
12 O x 16.0 g = <u>192 g</u>
342.3 g/mol

Step #4: For each element, take the mass of that element and divide it by the total mass of the compound.

% Al: $\frac{54\ g}{342.3\ g}$ = .1578

% S: $\frac{96.3\ g}{342.3\ g}$ = .2813

% O: $\frac{192\ g}{342.3\ g}$ = .5609

Note: I dropped the per mol part on the part where I'm calculating percent. By the end, the mass units should cancel, giving me only the percent sign as a unit.

Step #5: Multiply by 100 to turn this number into a percent.

% Al: $\frac{54\ g}{342.3\ g}$ = .1578 x 100 = 15.78%

% S: $\frac{96.3\ g}{342.3\ g}$ = .2813 x 100 = 28.13%

% O: $\frac{192\ g}{342.3\ g}$ = .5609 x 100 = 56.09%

Step #6: Check your answer by adding the percent values together. They should be very close to 100.
15.78% + 28.13% + 56.09% = 100%

Let's go the other way …
If you know the percent composition and the mass of a compound, you can calculate the contribution of a particular element to that compound.

Example: What mass of oxygen is contained in a 14.5 g sample of $Al_2(SO_4)_3$?
Answer: 14.5 g $Al_2(SO_4)_3$ x .5609 = 8.13305 g
To three sig figs, the answer is 8.13 g
Explanation: I'm using this compound since I just figured out the percent of oxygen in the compound. If 56.09% of the compound is oxygen, then it follows that oxygen should make up 56.09% of the mass in any sample. Therefore, I solve this by turning the percent into a decimal by dividing by 100 (undoing the percent) and multiply by the mass of the sample.

Empirical and Molecular Formulas:

Empirical formula: simplest form of a chemical formula; i.e. CH_2
Molecular formula: actual chemical formula for a compound; i.e. C_2H_4

Steps to finding empirical formulas:
Step #1: Find the # of moles for each element by doing a molar mass conversion to moles.

Note: If you're given percent values, then assume a 100 g sample. Therefore, your # of g = the %.

Step #2: Find the mole ratio of the elements to each other. If the numbers come out as a decimal, multiply them by a value that will make everything a whole number.
Step #3: Write the whole numbers from the ratio as the subscripts for each of the elements.

Example: Find the empirical formula of the chemical compound given that it contains 83.33% C and 16.67% H.
Step #1: Convert to moles.

83.33 g C x $\frac{1\ mol\ C}{12\ g\ C}$ = 6.94 mol C

16.67 g H x $\frac{1\ mol\ H}{1.0\ g\ H}$ = 16.67 mol H

Step #2: Find the mole ratio of the elements to each other. If the numbers come out as a decimal, multiply them by a value that will make everything a whole number.
Mole Ratio:
C : H
$\frac{6.94}{6.94} : \frac{16.67}{6.94}$
(1 : 2.4) x 5 = 5 : 12
Step #3: Write the whole numbers from the ratio as the subscripts for each of the elements.

Empirical formula: C_5H_{12}

Steps to calculating molecular formulas:

Step #1: Calculate the empirical formula.
Step #2: Calculate the formula mass/molar mass for the empirical formula.
Step #3: Divide the molecular formula mass/molar mass by the empirical mass to get a small whole number.
Step #4: Multiply the empirical formula by that small whole number value to get the molecular formula.

Example: A compound has a molar mass of 114 g/mol and a percent composition of 84.21% C and 15.79% H. Calculate the empirical and molecular formulas.
Step #1: Assume we worked through the steps above and got the empirical formula of: C_4H_9
Step #2: Calculate the molar mass for the empirical formula: (4 C x 12 g) + (9 H x 1.0 g) = 57 g/mol
Step #3: Divide the molecular mass by the empirical mass:
$\frac{114\frac{g}{mol}}{57\frac{g}{mol}} = 2$
Step #4: Multiply the empirical formula by the whole number to get the molecular formula: C_8H_{18}

Hydrates:

What is a hydrate?

In chemistry, it's a compound that has water as part of its formula.

Anhydrous crystal: a hydrate without the water; this is what you have left after burning off the water; that's a fun lab

Steps to finding moles of water in a hydrate:

Step #1: Find the mass of the anhydrous (without water) crystal and the water.
Step #2: Subtract the two to get the mass of the water.
Step #3: Find the moles of the anhydrous crystal and the moles of the water.
Step #4: Find the whole # ratio by dividing both mole amounts by the smallest amount of moles **Note:** The crystal amount is always smaller.
Step #5: Write the formula with a coefficient in front of the water to represent the amount of molecules contained in the crystal structure. Put a dot in between the salt and the water.

Example: 4.78 g of anhydrous lithium perchlorate was dissolved in water and re-crystalized. Care was taken to isolate all the lithium perchlorate as its hydrate. The mass of the hydrated salt obtained was 7.21 g. What hydrate is it?
Step #1: Find mass of substance without water. 4.78 g
Step #2: Find mass of water. 7.21 g - 4.78 g = 2.43 g
Step #3: Find moles of salt and water.

$4.78 \text{ g } LiClO_4 \text{ x } \frac{1 \text{ mol LiClO4}}{106.4 \text{ g LiClO4}} = 0.0449 \text{ mol}$

$2.43 \text{ g } H_2O \text{ x } \frac{1 \text{ mol H2O}}{18.0 \text{ g H2O}} = .135 \text{ mol } H_2O$

Step #4: Whole # ratio:

$LiClO_4\text{: } \frac{0.044929 \text{ mol}}{0.044929 \text{ mol}} = 1 \quad H_2O\text{: } \frac{0.13489 \text{ mol}}{0.044929 \text{ mol}} = 3$

Step #5: Write the formula. $LiClO_4 \cdot 3\ H_2O$

You can also find the formula of the hydrate if you know the percent composition of the compound. Let's use the previous problem as an example.

$LiClO_4 \cdot 3H_2O$ has a mass of 160.4 g of which 54 g is water. The water is 33.67% of the compound.
That means that the rest of the compound must be 66.33%. Then, you can find moles of each and set up a ratio.

Question: What is the formula for: $LiClO_4 \cdot n\ H_2O$
Step #1: Find mass of each. If you're given the percent of the water and the percent of the compound, then you can assume grams.

$$66.33\text{ g } LiClO_4 \text{ x } \frac{1\ mol\ LiClO4}{106.4\ g\ LiClO4} = .623\text{ mol } LiClO_4$$
$$33.67\text{ g } H_2O \text{ x } \frac{1\ mol\ H2O}{18\ g\ H2O} = 1.87\text{ mol } H_2O$$

Step #2: Set up a mole ratio.
$LiClO_4 : H_2O$
$$\frac{.623}{.623} : \frac{1.87}{.623}$$
1 : 3
Note: The part of the compound that's not water will always be the smaller number of moles.

Step #3: Write the empirical formula for the compound by plugging in the number of moles of water.
$LiClO_4 \cdot 3\ H_2O$

Most Common Mistakes:

- **Not using units!** Units are your friends. Allow them to help you convert from one substance to another.
- Units that don't match. If you want to cancel a unit, you must have the same unit (or unit/label combination)
  How many hours are in two years?
  2 years x $\frac{365\ days}{1\ year}$ x $\frac{24\ hours}{1\ day}$ = 17520 hours
- Improper abbreviations. Formula units = form u (no exceptions); moles = mol; molecules = molecules; atoms = atoms; grams = g; grams per mole = g/mol (not gm or gpm or whatever else your imaginations can come up with)

Chapter 8: Stoichiometry

Introduction:

If there's a single word in all of chemistry capable of striking fear into the hearts of students everywhere, "stoichiometry" would be it. In looking up the origins of the word, I found out that it's Greek for "element" with the slapped on English part of "metry." It happens to be one of my favorite parts of the course. Many of the concepts pop up elsewhere.

If you really want to know where the word came from, go here: http://www.etymonline.com/index.php?term=stoichiometry

Stoichiometry relates the parts of a balanced chemical equation to one another. If you start with a certain amount of reactants, stoichiometry will allow you to calculate how much of the product(s) you should be able to get from the reaction.

Understanding Mole Ratios:

One of the first exercises I have my students do is write out every mole ratio of a balanced chemical reaction. This simply gets them used to the idea that any part of a reaction can be related to any other part. For convenience, I assign the reactants and products letters, so they're easier to talk about in class. Combustion reactions are convenient to use in stoichiometry problems because they give us interesting mole ratios. There definitely could be a mole ratio of 1 to 1, but it's less interesting to speak about.

Example:
$C_3H_8 + 5\ O_2 \rightarrow 3\ CO_2 + 4\ H_2O$
A B C D
You can relate A to B, A to C, A to D, B to C, B to D, or C to D.

A to B: $\frac{1\ mole\ C3H8}{5\ moles\ O2}$	A to C: $\frac{1\ mole\ C3H8}{3\ moles\ CO2}$	A to D: $\frac{1\ mole\ C3H8}{4\ moles\ H2O}$
B to C: $\frac{5\ moles\ O2}{3\ moles\ CO2}$	B to D: $\frac{5\ moles\ O2}{4\ moles\ H2O}$	C to D: $\frac{3\ moles\ CO2}{4\ moles\ H2O}$

Note: The flipped versions of any of these are still valid mole ratios, but I think writing it out one way is good enough to get the point that they're related.

What is ideal stoichiometry?
How does it differ from limiting reactant stoichiometry?

Ideal stoichiometry (as opposed to limiting reactant stoichiometry) assumes perfect reactions. In other words, no loss of reactants and 100% yield for products. Limiting reactant stoichiometry assumes more realistic conditions where only some reactant molecules go through the chemical reaction and give you product. Limiting reactant stoichiometry also acknowledges that you don't usually start with exactly enough of each reactant. Typically, something runs out first and limits the amount of product that can be formed.

Four Types of Ideal Stoichiometry Problems:

There are four types of problems you will likely be asked to solve for in ideal stoichiometry: mole to mole, mole to mass, mass to mole, and mass to mass. The steps to solving these problems are the same as any other conversion problem. For each reaction, you need a balanced chemical equation. Let's continue to use the combustion of propane.

Review of Using Conversion Factors to Convert Units:
Step #1: Start with given information
Step #2: Put units for the unit that needs to change on the bottom and units that you're changing to on the top
Step #3: Repeat step 2 until you get to the final unit
Step #4: Plug in the conversion factors
Step #5: Solve (top = multiply, bottom = divide)

Mole to Mole Problems: one step

Example 1: $C_3H_8 + 5\ O_2 \rightarrow 3\ CO_2 + 4\ H_2O$
Question: If 3.15 moles of O_2 react, how many moles of CO_2 could be formed?

Answer: 3.15 moles O_2 x $\frac{3\ moles\ CO2}{5\ moles\ O2}$ = 1.89 moles CO_2

Explanation: Start with your given amount of moles of O_2, then set up a conversion factor. The mole ratio between oxygen and CO_2 is $\frac{5\ moles\ O2}{3\ moles\ CO2}$ or $\frac{3\ moles\ CO2}{5\ moles\ O2}$. Since you need to cancel oxygen, you need the second version of the mole ratio with the five moles of O_2 on the bottom and the three moles of CO_2 on the top. This is the only conversion factor needed at this time. Solve by multiplying by three and dividing by five. Report your answers as your teacher requires. Most will say to go to the same sig figs as the given amount. 3.15 has three sig figs, so the answer would be 1.89 moles CO_2.

Mole to Mass Problems: two steps

Example 2: $C_3H_8 + 5\ O_2 \rightarrow 3\ CO_2 + 4\ H_2O$
Question: If 3.15 moles of O_2 react, what mass of CO_2 could be formed?

Answer: 3.15 moles O_2 x $\frac{3\ moles\ CO2}{5\ moles\ O2}$ x $\frac{44\ g\ CO2}{1\ mol\ CO2}$ = 83.16 g CO_2 or 83.2 g CO_2

Explanation: The problem starts out exactly the same as before by using the mole ratio of $\frac{3\ moles\ CO2}{5\ moles\ O2}$ to convert out of moles of O_2 and into moles of CO_2. Since the problem asks for the grams of CO_2, you need the molar mass of carbon dioxide to convert out of moles of CO_2 and into grams of CO_2. Note that the mole ratio of CO_2 could be written $\frac{1\ mol\ CO2}{44\ g\ CO2}$ or $\frac{44\ g\ CO2}{1\ mol\ CO2}$. Because you're already in moles of CO_2 by the end of the mole ratio step, you need the second version of the molar mass conversion factor with the grams on top. Solve as usual by multiplying across the top and dividing across the bottom.

Tip: It shouldn't matter which order you do the steps in as long as you do everything.

Mass to Mole Problems: two steps

Example 3: $C_3H_8 + 5\ O_2 \rightarrow 3\ CO_2 + 4\ H_2O$

Question: In an ideal reaction, how many moles of water (H_2O) could be formed from a reaction of 74.5 g O_2?

Answer: 74.5 g O_2 x $\frac{1\ mole\ O2}{32\ g\ O2}$ x $\frac{4\ moles\ H2O}{5\ moles\ O2}$ = 1.8625 moles H_2O or 1.86 moles H_2O

Explanation: Start with the given of 74.5 g O_2. The first conversion factor is a molar mass to get out of grams of O_2 and into moles of O_2 because you can't do a mole ratio until you're in the unit of moles. Once you're in moles of O_2, do another conversion from moles of O_2 to moles of H_2O. This step requires the mole ratio of oxygen to water. Of the two possible versions of the conversion factor, you need the one with the moles of water on top because that is the unit you were asked to switch to. Solve by multiplying across the top and dividing anything on the bottom. To three sig figs, the answer is 1.86 moles H_2O.

Mass to Mass Problems: three steps

Example 4: $C_3H_8 + 5\ O_2 \rightarrow 3\ CO_2 + 4\ H_2O$

Question: In an ideal reaction, what mass of water (H_2O) in grams could be formed from a reaction of 74.5 g O_2?

Answer: 74.5 g O_2 x $\frac{1\ mole\ O2}{32\ g\ O2}$ x $\frac{4\ moles\ H2O}{5\ moles\ O2}$ x $\frac{18\ g\ H2O}{1\ mol\ H2O}$ = 33.525 g H_2O or 33.5 g H_2O

Explanation: This problem is identical to the one before except for the tiny detail of the end unit. This time, you're starting out with 74.5 g O_2 and asked how many grams of water could be formed. The solution involves the exact same two steps in the previous example and one more conversion. Instead of ending with moles of water, use the molar mass of water to convert to grams. At the end of the last problem, you were in moles H_2O. To convert out of moles of water and into grams of water, you need to multiply by 18 g H_2O. The final answer to three sig figs is 33.5 g H_2O.

Note: Your teacher might be particular about how you show these conversion factors, or they may teach you another way entirely. I've seen stoichiometry done with ratios too. It works, but I think this method has a better chance of letting students see where they started and watch the progress through the problem until there's a solution. I require all my students to write a number, unit, and label on every step. It's somewhat tedious, but it's also the best way of tracking where you came from and where you're going.

Limiting Reactant Stoichiometry: What if something runs out first?

In most chemical reactions, something is going to be consumed first. This is the limiting reactant. It's the reactant that determines how much of each product should be created.

Note: You may see the term "limiting reagent." These terms are used interchangeably, but there are slight differences in the definitions. Typically, you're safe enough if you stick with LR, which covers both.

When going through this part of the unit, I ask five questions for every problem set.
1. What is the limiting reactant (LR)?
2. How many grams of the excess reactant are left over?
3. How many grams of each product could be produced in an ideal reaction?
4. If ___ g of (insert name of one product) were actually produced, what is the percent yield?
5. Assuming the same percent yield, how much of the other product would actually be produced?

Example: $C_3H_8 + 5\ O_2 \rightarrow 3\ CO_2 + 4\ H_2O$
Question: If 135 g of each reactant were present …
1. What is the limiting reactant (LR)?

Two Methods for Finding Limiting Reactants:

Method #1: Choose one of the reactants, assume it's the LR, change to the other reactant, and evaluate. If you have more than you need, you guessed the LR correctly. If you need more than you have, you guessed incorrectly, the other reactant is the LR.

$$135\text{ g } C_3H_8 \times \frac{1\ mol\ C3H8}{44\ g\ C3H8} \times \frac{5\ mol\ O2}{1\ mol\ C3H8} \times \frac{32\ g\ O2}{1\ mol\ O2} = 491\text{ g } O_2$$

Analysis and conclusion: The key question is: do I have what I need or do I need more than I have? I've made an assumption that C_3H_8 is the limiting reactant and asked how many grams of O_2 I need to react at the same time. I was given 135 g of each reactant, but in order for C_3H_8 to be the LR, I need 491 g of O_2. Therefore, I guessed wrong. The LR for this reaction must be O_2.

Note: It's a complete, annoying coincidence that the molar mass of C_3H_8 and CO_2 are the same.

$C_3H_8 + 5\ O_2 \rightarrow 3\ CO_2 + 4\ H_2O$

Method #2: Start with both reactants, assume both are LR, go to the SAME product, and evaluate.

$$135\ g\ C_3H_8\ x\ \frac{1\ mol\ C3H8}{44\ g\ C3H8}\ x\ \frac{4\ mol\ H2O}{1\ mol\ C3H8}\ x\ \frac{18\ g\ H2O}{1\ mol\ H2O} = 221\ g\ H_2O$$

$$135\ g\ O_2\ x\ \frac{1\ mol\ O2}{32\ g\ O2}\ x\ \frac{4\ mol\ H2O}{5\ mol\ O2}\ x\ \frac{18\ g\ H2O}{1\ mol\ H2O} = 60.8\ g\ H_2O$$

Analysis and conclusion: The reactant that gives less of the chosen product is the LR. In this example, the O_2 resulted in fewer grams of water, therefore it must be the LR.

Caution: Do not go to different products! Your goal is to compare the amounts of the products formed. This does not work if you go to different products.

Note: You can stop after moles in order to answer the question. I tend to go that extra step because I know the third question is going to be mass of products anyway. The set of conversions that involves the LR is now half of that answer.

Commentary: If given a choice, I go with Method 2 because I'm more comfortable with that evaluation. However, Method 2 has a very common pitfall: people often circle the lower amount of product and call it the LR. The LR is a reactant; therefore, it can never be a product. It's the reactant that resulted in the smaller amount of product. Method 1 also works just fine, but when you guess incorrectly, the analysis can look confusing.

$C_3H_8 + 5\ O_2 \rightarrow 3\ CO_2 + 4\ H_2O$

2. How many grams of the excess reactant are left over? (Start with the LR and change to the other reactant. Then, subtract this amount from the available starting amount of that other reactant.)

$135\ g\ O_2 \times \frac{1\ mol\ O2}{32\ g\ O2} \times \frac{1\ mol\ C3H8}{5\ mol\ O2} \times \frac{44\ g\ C3H8}{1\ mol\ C3H8} = 37.1\ g\ C_3H_8$ used in the reaction

Available – used = excess left over

$135\ g\ C_3H_8 - 37.1\ g\ C_3H_8 = 97.9\ g\ C_3H_8$ left over

Explanation: Start with the limiting reactant and change completely over to non-limiting reactant. This tells you how many grams of the non-LR you use during the course of the chemical reaction. A lot of students stop there, but there is one more step. You've not quite answered the question yet. In order to find out how many grams are left over, you need to subtract the answer you just got from the amount of starting non-limiting reactant you were given.

$C_3H_8 + 5\ O_2 \rightarrow 3\ CO_2 + 4\ H_2O$

3. How many grams of each product could be produced in an ideal reaction? (Start with the LR and change over to each of the products in turn. These are your theoretical yields. That means you assume 100% of the limiting reactant results in product. It's the maximum amount you could produce from the LR.)

$135\ g\ O_2 \times \frac{1\ mol\ O2}{32\ g\ O2} \times \frac{3\ mol\ CO2}{5\ mol\ O2} \times \frac{44\ g\ CO2}{1\ mol\ CO2} = 111.\ g\ CO_2$

$135\ g\ O_2 \times \frac{1\ mol\ O2}{32\ g\ O2} \times \frac{4\ mol\ H2O}{5\ mol\ O2} \times \frac{18\ g\ H2O}{1\ mol\ H2O} = 60.8\ g\ H_2O$

4. If 93.4731 g of CO_2 were actually produced, what is the percent yield? (answer to two decimals)

Percent yield $= \frac{actual\ yield}{theoretical\ yield} \times 100$

Percent yield $= \frac{93.4731\ g\ CO2}{111\ \ g\ CO2} \times 100 = 84.21\%$

Explanation: Any percent is a part of something over the whole multiplied by 100. In this question, the grams of the CO_2 are the "actual yield." The theoretical yield comes from the previous question. Multiplying by 100 simply turns the number into a percent.

Caution: It's easy to flip the numbers, but always evaluate your answer. Ask yourself: could this happen? The answer cannot be above 100% and probably shouldn't be something crazy small either. (That last one might be a personal bias, but I tell my students, when I make up the questions, I typically don't give things less than 50% yield. If you're getting .22%, it's probably wrong.)

5. Assuming the same percent yield, how much of the other product would actually be produced? (Turn the percent yield back into a decimal by dividing by 100 and multiply by the other actual yield.)

$$\text{Actual yield} = \frac{percent\ yield\ x\ theoretical\ yield}{100} =$$

$$\text{Actual yield} = \frac{60.8\ \text{g H2O}\ x\ 84.21\%}{100} = 51.2\ \text{g}\ H_2O$$

Chapter 9: The Kinetic Molecular Theory and Gases

Introduction:

As far as I know, this is one unit being phased out by the Next Generation Science Standards. However, because NGSS hasn't been adopted everywhere and in the interest of being thorough, I'm still going to include it here. I've always liked the gases unit because it gives a few openings to talk about real world applications.

Kinetic Molecular Theory of Gases:

This theory has been presented in detail on numerous websites. It describes gases as consisting of many small particles with a lot of empty space between them that don't affect each other unless they collide. These particles move randomly in a straight line unless they hit other gas particles or a container wall. The temperature and the speed of the particles are inextricably linked. The next two points are where you realize this theory speaks only of ideal gases, which don't exist. However, real gases behave just like imaginary (ideal) ones as long as the pressure's not too high and the temperature's not too low. The points I'm referring to are where the Kinetic Molecular Theory speaks of gases being able to collide without losing any energy and gas particles not interacting with each other or the container.

If you'd like to know more, check out this site: http://chemed.chem.purdue.edu/genchem/topicreview/bp/ch4/kinetic4.html

Why should you care?

Reason #1: Understanding the gas laws can keep you safe. Why shouldn't you throw an aerosol can into a fire? Why shouldn't you immediately rise to the surface if you've been deep sea diving?

Reason #2: Gas laws can explain some of adults' weird behavior. Why do people put more air into tires during cold winter months? Why do people use pressure cookers, especially if they have to cook things at high altitudes?

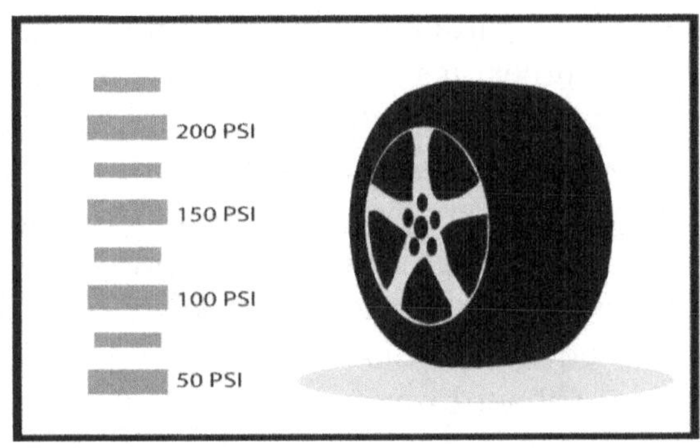

Reason #3: Gas laws can explain some interesting recreational events and other randomness. How do hot air balloons work? How can you crush a can with the power of your mind and some air pressure? Why is soda fizzy? How much gas fills up a deploying airbag in a car? What happens to a marshmallow that's put in a microwave? Can water be boiled at room temperature? Why do your ears pop on an airplane?

What is Air Pressure?

Pressure caused by air. I can feel the evil stares through time and space. Air is composed of gas particles. You can look up the exact mixture of atmospheric gas, but it's mostly nitrogen with a good chunk of oxygen and then some other gases mixed in. Anyway, air pressure is caused by the weight of gaseous air molecules. This can be measured with an instrument like a barometer.

Why does it take longer to cook an egg on a mountain?

Here's your answer: air pressure. As altitude increases, air pressure decreases. Since air pressure is tied to boiling point, this explains why it takes longer to cook something at very high altitudes compared to sea level. Boiling happens when the internal pressure of the liquid is equal to the external pressure. So if that external pressure is air pressure and it's lower than it would be at sea level, the boiling point will also be lower than it would be at sea level. If the boiling point is lower, the cooking temperature of the thing boiling is lower and will therefore take longer overall.

How does a pressure cooker work?
Chemistry teachers are also fond of that question. The answer involves the same concept, opposite point as the "eggs on a mountain" question. A pressure cooker will increase the pressure over the liquid, which causes a delay in the boiling point. When the boiling point is raised, the substance ends up cooking at a higher temperature. If something's cooking at a higher temperature, it will ultimately cook faster.

The useful and probably most frustrating part about air pressure will be how many different units you can measure it in. I typically stick to atmospheres (atm) and millimeters of mercury (mm Hg). But just to confuse chemistry students mm Hg can also be referred to as torr after Torricelli, the guy who did a lot of work studying air pressure. As if that wasn't confusing enough, there's also pounds per square inch (psi), bar, barye (Ba), pascals (Pa), and kilopascals (kPa).

Want to crush a can?
Don't do this without adult supervision and permission. Get an empty, clean soda can and put a small amount of water in the bottom of it. Then, heat it up on a stove until there's a good amount of steam coming out. When it looks like it's stopped steaming, use hot hands to grasp the can and flip it upside down into a container of cold water so that the opening is sealed. If you've done it correctly, the air pressure should be enough to crush the can.

Air pressure's pretty powerful. It's pressing in around us all the time. So, why aren't we crushed? The simplistic answer is that there's stuff in us. Normal cans aren't crushed by air pressure either, but that's because there's air pressure inside the can pressing out. By putting in water and letting it steam, you carried out that air.

The Gas Laws:

Three things are linked when it comes to gases: pressure, volume, and temperature. There are three basic laws that will address those relationships: Boyle's Law, Charles's Law, and Gay-Lussac's law. Together, you'd get the combined gas law. The amount of a gas also matters. The ideal or universal gas law deals with that relationship. There's also Avogadro's Law, which allows you to do gas stoichiometry if a gas is held at standard temperature and pressure conditions.

Useful conversion factors:
760 torr/mmHg = 1 atm
°C + 273.15 = K
1000 mL = 1 L

Note: All gas law problems need to be worked out in Kelvin temperatures. The Celsius scale is based off of the behavior of water. The temperature water freezes is called 0°C and the temperature where water boils is called 100°C.

Boyle's Law (Pressure and Volume): $P_1V_1 = P_2V_2$
Boyle's Law features pressure and volume and shows the inverse or indirect relationship that exists between the two. Temperature remains constant. Pressure is caused by molecules hitting the walls of their container. Gases completely fill their container so that their volume matches whatever container they're placed in.

- If the volume is small, there's not much room for the molecules and the frequency with which they hit the walls of the container will be higher. Therefore, the pressure will be high.

- If the volume is large, there's a lot of room for the molecules and the frequency with which they hit the walls of the container will be low. Therefore, the pressure will be low.

So, if volume increases, pressure decreases. If volume decreases, pressure increases. That's what is meant by saying that pressure and volume are inversely related to each other.

Example: If a gas in a 2.50 liter (L) container has a pressure of 4.00 atm, what is the new pressure when the gas is moved to a 5.00 L container?

Answer: I prefer setting up a chart. It takes an extra second but helps me organize my thoughts.
$P_1 = 4.00$ atm $P_2 = ?$
$V_1 = 2.50$ L $V_2 = 5.00$ L

Next, I write down the formula and rearrange it. My goal is to algebraically solve the equation for the unknown piece of information.
$P_1V_1 = P_2V_2 \quad P_2 = \frac{P1V1}{V2}$
After that, I plug into the rearranged formula and solve.
$P_2 = \frac{P1V1}{V2} \qquad P_2 = \frac{(4.00\ atm)(2.50\ L)}{(5.00\ L)} = 2.00$ atm

Finally, I check the answer to see if it makes sense.
The volume has gotten bigger, so the pressure should drop. Since the volume doubled, it follows that the pressure is cut in half.

Charles's Law (Volume and Temperature): $V_1T_2 = V_2T_1$
The volume and the absolute temperature of a gas are directly linked. The term "absolute temperature" simply means you can't use the Celsius scale, you have to use Kelvin. Pressure remains constant. If the pressure's to remain constant, gas particles at higher temperatures will be taking up more space and the volume gets bigger accordingly.

Note: You may see other versions of Charles's Law, such as:
$\frac{V1}{T1} = \frac{V2}{T2}$

The version I have written above is simply a reworked version that makes it easier for you to solve for the unknown. At this point, you isolate the variable (thing) you're solving for by dividing the thing it's next to. For example, if you want to solve for V_1, divide both sides by T_2.

Example: What was the original volume of gas if it started at a temperature of 305 K and was cooled to 245 K? The final volume was 6.77 L.

Answer: Once again, I take the time to set up a chart of the known information.

$T_1 = 305$ K $\quad T_2 = 245$ K
$V_1 = ?$ $\quad V_2 = 6.77$ L

Then, I write down the formula and rearrange it to solve for V_1.
$V_1T_2 = V_2T_1 \quad V_1 = \frac{V2T1}{T2}$

Next, I plug in the information I have and solve.
$V_1 = \frac{V2T1}{T2} \qquad V_1 = \frac{(6.77\ L)(305\ K)}{(245\ K)} = 8.43$ L

Finally, I evaluate the answer.

The temperature is dropping so the volume should also be dropping, but I'm solving for the first volume. Therefore, the first volume (my answer) should be bigger than the second volume which was given. 8.43 L is bigger than 6.77 L, so the answer makes logical sense.

Gay-Lussac's Law (Pressure and Temperature): $P_1T_2 = P_2T_1$
The pressure and the temperature of a gas are also directly linked. Volume is held constant. Pressure is caused by the molecules colliding with the walls of their container. Molecules that are moving at faster speeds are more likely to collide more often, so the pressure will go up as the temperature goes up. The opposite is also true. Molecules that are cooled down are slower and collide with the walls less often, so the pressure drops.

Example: What is the new temperature if gas held under a pressure of 3.50 atm starting at 445 K has its pressure increased to 9.48 atm?

Answer: The chart would look like this:
$T_1 = 445$ K $\quad T_2 = ?$
$P_1 = 3.50$ atm $\quad P_2 = 9.48$ atm

The original formula of $P_1T_2 = P_2T_1$ can be rewritten to $T_2 = \frac{P2T1}{P1}$.

When I plug in the information I was given …
$T_2 = \frac{P2T1}{P1} \quad T_2 = \frac{(9.48\ atm)(445\ K)}{(3.50\ \text{atm})} = 1205.31429$ K or 1210 K

Finally, I evaluate the answer. The pressure has increased, so the temperature should also increase.

Combined Gas Law: $P_1V_1T_2 = P_2V_2T_1$
You might see it written as $\frac{P1V1}{T1} = \frac{P2V2}{T2}$. I prefer the version that's written as one line because it's easier to solve for an unknown by dividing both sides by the other two variables.

There's no easy way to predict the answer to a question involving all six variables, but a chart will help you keep organized. If you don't want to write out an official chart, at the very least jot down what each piece of information is in a problem. The question has to give you enough information.

Incidentally, there's a card you can make that will aid with deciding which of the gas laws you're trying to solve. It's easier just to show you, but hard to type out. I will do the best I can, but I'll also try to upload a sample on to my website.

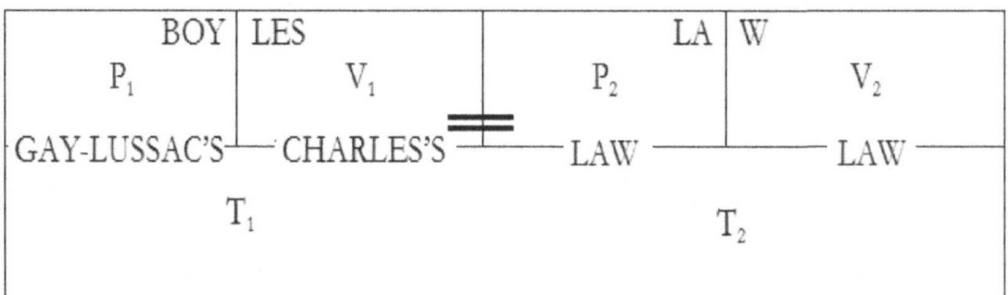

Explanation of card:
Take a notecard and fold it in half the long way like a hot dog. Then, fold each side in to the center and crease the top half. Then, write the cards as you see them above. "Gay-Lussac's" and "Charles's" as well as both bottom "law" must be written on the long crease itself. The shorter creases should be cut with scissors between the word "Boyles" and the top "law."

Once prepared, fold back the thing held constant. So, if temperature is constant, only p's and v's should be showing. The only words you should be able to read in full would be Boyle's Law. What you see is the formula for Boyle's Law. If the volume is held constant, P's and T's should be visible. The only words you should be able to read in full should be Gay-Lussac's Law.

Note: These are the original versions of the laws, not the rewritten versions. To generate the versions I used before bring the T's up to the opposite side. So T_2 would stand next to P_1 and T_1 would stand next to P_2.

Avogadro's Law: 22.4 L = 1 mole at STP

Under standard conditions, a mole of gas will occupy 22.4 L of space. This relationship allows you to do gas stoichiometry problems.

Standard temperature and pressure (STP):
1 atm and 273.15 K.

Let's return to the equation we were using before to do stoichiometry problems. Assume the propane's a gas. In fact, assume everything's a gas in this equation.

$C_3H_8 + 5\ O_2 \rightarrow 3\ CO_2 + 4\ H_2O$

Example: If 3.56 L of oxygen reacts with excess propane, what mass of water could be produced?

$$3.56\ L\ O_2 \times \frac{1\ mol\ O2}{22.4\ L\ O2} \times \frac{4\ mol\ H2O}{5\ mol\ O2} \times \frac{18\ g\ H2O}{1\ mol\ H2O} = 2.29\ g\ H_2O$$

Explanation: Start with the given L of oxygen and convert to moles of O_2 by dividing by 22.4 L. This is an application of Avogadro's Law. Next, convert out of moles of O_2 and into moles of H_2O by using the mole ratio. Then, convert to grams of H_2O by multiplying by the molar mass of water, which is 18 g of H_2O per mole of H_2O.

Ideal Gas Law (Universal Gas Law): PV = nRT

This law allows for calculations with moles of gas that are not at standard pressure and temperature conditions. Occasionally, stoichiometry is required to solve these problems because we're not always given the moles.

Here is the ideal gas law solved for each variable. You will not be asked to solve for R because that's a constant. There are several R values. The most popular one being .0821 $\frac{L\,atm}{mol\,K}$. The only difference in the R values is the unit for pressure and volume. You must match your units to the R value you intend to use. This may require you to do several conversions before plugging into the equation.

$PV = nRT$

$P = \frac{nRT}{V}$ $\quad V = \frac{nRT}{P}$

$n = \frac{PV}{RT}$ $\quad T = \frac{PV}{nR}$

Example: $C_3H_8 + 5\ O_2 \rightarrow 3\ CO_2 + 4\ H_2O$
If 5.87 moles of O_2 react with excess propane, what volume will the resulting CO_2 gas occupy if the temperature is 358 K and the pressure is 6.52 atm?

First, set up a chart:
P = 6.52 atm V = ? n = 3.522 mol CO_2
R = .0821 $\frac{L\,atm}{mol\,K}$ T = 358 K

To find the n, you need to do stoichiometry.
5.87 mol O_2 x $\frac{3\ mol\ CO2}{5\ mol\ O2}$ = 3.522 mol CO_2

Solve the ideal gas law for V because that is the only unknown.
$PV = nRT$ $\quad V = \frac{nRT}{P}$

Plug in the information you gathered.
$PV = nRT$ $\quad V = \frac{nRT}{P} = \frac{(3.522\ mol\ CO2)(.0821\frac{Latm}{molK})(358\ K)}{6.52\ atm} = 15.9\ L$

Caution: If there are two variables on the bottom of the fraction, be careful when you put the information in your calculator. It's a very common mistake to try to do too much at once without proper parentheses. In such cases, the calculator then ends up multiplying by the last number, which it should have divided by.

Effusion and Diffusion:

Diffusion is the idea that particles, gases in this case, move from an area of high concentration to low concentration.

Effusion is a scenario where you place different gases in closed containers, poke holes into those containers, and try to figure out which one would escape first. The key to that is the idea that tiny gases move faster than bigger gases. So if one container holds H_2 and another holds an equivalent amount of Cl_2, the H_2 will effuse faster.

Answer to some of those burning questions brought up before.

I believe some were answered along the way, so I'll skip those.

Question: Why shouldn't you throw an aerosol can into a fire?

Answer: Gay-Lussac's Law. The pressure and temperature are directly linked. If you throw a can into a fire, the pressure will build up, and likely, the can will explode, which can't be healthy for anybody standing near that can.

Question: Why shouldn't you immediately rise to the surface if you've been deep sea diving?

Answer: Boyle's Law. The pressure and the volume are inversely related. If you rise too quickly, the volume of the gases in your blood will expand as well. This leads to the bends, which means a lot of pain, and in extreme cases, death.

Question: Why do people put more air into tires during cold winter months?

Answer: Gay-Lussac's Law. If the outside temperature is low, the particles of air within the tires will be moving slower, so the pressure will drop. In newer cars, this means angry lights on the dashboard yelling at you to put more air in your tires to compensate for the lower air pressure.

Question: How do hot air balloons work?
Answer: Charles's Law. Heating the air inside the balloon causes it to expand. The extra volume trapped by the balloon will allow the warm air inside to be less dense than the air outside the balloon. When that happens, the balloon rises.

Question: Why is soda fizzy?
Answer: When soda is bottled, it's done at a pressure greater than 1 atm. When you pop a can of soda open or untwist the cap, the pressure returns to 1 atm. Gases are more soluble under higher pressures, so when you return it to normal, some of the extra CO_2 gas is forced out of the solution. It escapes as bubbles.

Question: How much gas fills up a deploying airbag in a car?
Answer: You'd have to know the balanced chemical equation for the reaction of chemicals put into your airbag as well as how much of that chemical is reacting and the pressure and temperature conditions. Then, questions can be asked and answered using the ideal gas law.

Question: What happens to a marshmallow that's put in a microwave?
Answer: Charles's Law. Marshmallows are mostly air. If you put a marshmallow in a microwave, the air inside and around it gets hot. As the air gets hot, it will expand, making the marshmallow bigger. I buy marshmallow Peeps for the sole purpose of microwaving them.

Question: Can water be boiled at room temperature?
Answer: Yes, lower the air pressure around the water. Boiling occurs when the atmospheric pressure equals the internal pressure of the liquid. If you lower the pressure, the temperature you need to reach to boil something will drop.

Chapter 10: Solutions

Introduction:

The chapter on solutions contains a lot of vocabulary terms. In order to discuss solutions, you need to be familiar with those definitions. The math involved in the unit deals with various ways to describe concentration.

Definitions:

Solution: a homogenous mixture where one type of particle is evenly spread throughout another
Solute: the smaller component in a solution; the thing added
Solvent: the larger component in a solution; the thing doing the dissolving; water is the universal solvent
Miscible: when two liquids are able to be dissolved in one another, forming a solution, they are said to be miscible
Immiscible: when two liquids are unable to be dissolved in one another and remain as separate entities
Solubility: the ability for something to dissolve in a different substance, forming a solution; solubility is particular to temperature and pressure conditions
Rate of solubility: how quickly something dissolves in something else

Like Dissolves Like:

This refers to the fact that polar solutes will dissolve in polar solvents and nonpolar solutes will dissolve in nonpolar solvents. Polarity refers to whether or not the molecule has "poles," a positive and negative side. For example, water's a very polar molecule. As a solvent, it will dissolve other polar molecules or ionic compounds that break down into their ions. Oil, which is very nonpolar, will not be able to dissolve in water. This is why when you add oil to water you will see two layers form.

Factors that increase rate of solubility:

In order for a solute to dissolve in a solvent, the two particle types need to meet. Therefore, anything that increases the rate at which solvent particles meet solute particles will increase how quickly the solution is formed.

- Heating the solvent
- Stirring or shaking
- Increase the surface area of the solute

Reading Solubility Curves:

You may be asked to read a solubility curve. They're a tad intimidating because they contain a lot of information. Along the left side, you'll see amounts, usually in grams per 100 g of water. This is how much of a particular substance should be able to be dissolved in 100 g of water at a particular temperature. The temperatures will be written along the bottom, usually in °C.

When given a solubility problem, find the temperature along the bottom and the amount along the side and see where it meets the correct chemical. You may be asked to simply report a certain value, compare it to a different value, or to manipulate it in some way.

With most substances, as the temperature goes up, the solvent molecules will be moving around more and the solubility of the substance will increase. This means that you should be able to dissolve more of the solute in that solvent at higher temperatures. This is not true for gases. Gas solubility goes down with increasing temperature because the molecules gain enough energy to escape the liquid phase and go into the gas phase.

Saturated, Unsaturated, Supersaturated:

Solubility, that is the amount of solute that can be dissolved in a given amount of solvent, is dependent upon certain temperature and pressure conditions. If a solution is saturated, it's holding the maximum amount of solute that it should be able to at that temperature and pressure. In many cases, solubility can be increased by increasing the temperature of the solvent.

If a solution is unsaturated, you should be able to add more solute to the solution and have it dissolve. If a solution is supersaturated, it is already holding more solute than it should be able to at that temperature and pressure. Certain substances like sodium acetate can be supersaturated. Basically, you saturate them at a high temperature and slowly cool them down. If you do it right, the extra solute will remain in solution even after you cool it down. Therefore, it's supersaturated because it's holding more solute than it should be able to under normal conditions.

How can you tell the types of solutions apart?

Add a little more solute to it and see how it behaves. If the solution is unsaturated, the additional solute should simply dissolve, so you would see no change in the solution. If the solution is saturated, the additional solute should drop to the bottom, so you should be able to see it. If the solution is supersaturated, the entire solution should crystalize.

Electrolytes vs. Nonelectrolytes:

An electrolyte is a substance that will break down in water, resulting in a solution that conducts an electric current. The key to this is that you need ions to carry the current on. The more ions that are present, the easier it is to conduct the current.

A nonelectrolyte is a substance that will break down in water, resulting in a solution that cannot conduct an electric current. Something like sugar would be a non-electrolyte. Ionic substances that are insoluble in water are also nonelectrolytes because their ions are not available to conduct electricity.

Colligative Properties:

These are properties like the freezing point and the boiling point that depend on the amount of particles present, not the identity of those particles. You might be asked to solve freezing point depression or boiling point elevation problems. The math looks more complicated than it really is, but I'm going to spend my limited time here talking you through the concepts.

Essentially, the more particles you have, the more stable the liquid state will be. This leads to a higher boiling point than usual and a lower freezing point than usual. Have you ever wondered why somebody threw salt into water they want to boil or why people "salt" their driveways in the winter? The answer lies in the colligative properties.

Adding salt to water delays the boiling point. When table salt ($NaCl$) is added to water, it breaks down into Na^{+1} and Cl^{-1} ions, putting more particles in the water. If you add enough extra particles, the boiling point will be delayed slightly. So, once the water's finally boiling, it will be at a hotter temperature than normal boiling. Therefore, whatever you're cooking should finish faster.

Adding salt to a walkway or road lowers the freezing point. Often times, this is enough to allow wet roadways to not become covered in ice. Road salt is typically something like $CaCl_2$ as opposed to $NaCl$ because you get more particles. Breaking down a molecule of $CaCl_2$ results in three particles: one Ca^{+2} ion and two Cl^{-1} ions. Breaking down a molecule of $NaCl$ results in two particles: one Na^{+1} ion and one Cl^{-1} ion. Incidentally, table salt is capable of lowering the freezing point, it's just not as effective as

$CaCl_2$.

Molarity, Molality, Mass Percent, and Dilution:

Molarity and molality are two different ways of expressing concentration. Molarity (M) is the moles of solute per liter of solution. Molality (m) is the moles of solute per kilogram of solvent. The only way to complicate these types of problems is to give you the information in other units so that you have to do conversions before plugging into the formula.

Conversions:
1000 mL = 1 L
1000 g = 1 kg

Molarity formula:

$M = \frac{moles\ solute}{L\ solution}$ moles solute = M x L solution

L solution = $\frac{moles\ solute}{M}$

Example 1: What is the Molarity of a solution if 2.00 L of solution contain 6.00 moles of NaCl?
Determine which version of the formula you need.

$M = \frac{moles\ solute}{L\ solution}$

Convert to the correct units and plug in.
$M = \frac{6.00\ mol\ NaCl}{2.00\ L}$ = 3.00 mol/L or M

Note: The unit for molarity can be a capital M or it can be moles over liters.

Example #2: What is the Molarity of a solution if 345 g of NaCl is dissolved in enough water to make 6252 mL of solution?

This problem is more complicated because you're not immediately given the units you need for the formula.

Determine which version of the formula you need.

M = $\frac{moles\ solute}{L\ solution}$

Convert to the correct units.

345 g NaCl x $\frac{1\ mol\ NaCl}{58.5\ g\ NaCl}$ = 5.90 mol NaCl

6252 mL x $\frac{1\ L}{1000\ mL}$ = 6.252 L

Plug into the formula.

M = $\frac{5.90\ mol\ NaCl}{6.252\ L}$ = .944 M

Molality formula:

m = $\frac{moles\ solute}{kg\ solvent}$ moles solute = m x kg solvent

kg solvent = $\frac{moles\ solute}{m}$

Example 3: How many grams of NaCl are needed to make .657 m solution containing .748 kg water?
Determine which version of the formula you need.
Moles solute = m x kg solvent

Convert to the correct units and plug in.
Moles solute = .657 m x .748 kg = .491 mol NaCl
Check the answer. .491 mol NaCl isn't the final answer. You need to convert to grams.

.491 mol NaCl x $\frac{58.5\ g\ NaCl}{1\ mol\ NaCl}$ = 28.7 g NaCl

Mass Percent:
Yet another way to present how much solute you have in a solution.

Mass % = $\frac{mass\ of\ solute}{mass\ of\ solution}$ x 100

Mass % = $\frac{mass\ of\ solute}{mass\ of\ solute + mass\ of\ solvent}$ x 100

Note: The unit of mass for the solute and solution must match, but it doesn't really matter what they are because they will cancel.

Most common mistake: People tend to forget that the solution includes the solute.

Example 4: What is the mass percent of a solution that contains 32.2 g of NaCl in 244 g H_2O?
The solute is the thing put into the solvent. It's there in lesser quantity. In this problem, the NaCl is the solute. The water is the solvent.

Mass % = $\frac{mass\ of\ solute}{mass\ of\ solution}$ x 100

Mass % = $\frac{32.2\ g\ NaCl}{32.2\ g\ NaCl + 244\ g\ H2O}$ x 100

Mass % = $\frac{32.2\ g\ NaCl}{276.2\ g\ solution}$ x 100   Mass % = 11.66%

Note: I tend to take all percentages to two decimals, but you should follow the instructions given by your teacher.

Dilution:
There are several versions of the dilution formula and the way it's presented.

$M_{concentrated} V_{concentrated} = M_{diluted} V_{diluted}$

$M_1V_1 = M_2V_2$

You dilute a solution by adding water to a concentrated solution so that the concentration of solute drops. The hardest part of doing these problems is correctly identifying which piece of information is which. Keep two things in mind: the second molarity (concentration) will always be lower than the first and the second volume will always be greater than the first. You add water to decrease the concentration.

Creating diluted solutions is probably a skill your teacher will need to use at some point to set up for a lab. Acids and bases tend to be sold in high concentrations.

Example 5: If a lab calls for 500 mL of a .5 M HCl solution, how many mL of 18 M HCl will be needed?
Since one of the hardest parts is determining which piece of information is which, I set up a chart.
$M_1 = 18$ M $\quad V_1 = ?$
$M_2 = .5$ M $\quad V_2 = 500$ mL

Next, I solve the original equation for the variable I'm looking for.

$M_1V_1 = M_2V_2 \qquad V_1 = \frac{M2V2}{M1}$

I plug in the given information.

$V_1 = \frac{(.5\ M)(500\ mL)}{(18\ M)} = 13.9$ mL

Safety note:

Do not add water to a concentrated acid or base. You're likely to cause it to splash that way. Concentrated acids or bases can easily cause severe burns. Besides taking proper precautions like wearing goggles, a chemical resistant apron, and gloves, you should pour the concentrated acid into a smaller, more manageable container that's easier to pour from, like a small beaker. Then, pour it down a glass stir rod into the water you're diluting it with.

Solution Stoichiometry:

Molarity can be used as a conversion factor, allowing you to solve stoichiometry problems. Remember, molarity is moles over liters.

Example 6: 2 HNO_3 + $Ca(OH)_2$ → $Ca(NO_3)_2$ + 2 H_2O
If .245 L of .784 M HNO_3 reacts with .541 L of 1.79 M $Ca(OH)_2$, how many g of water could be produced?
Because the problem gives two starting amounts, this is a limiting reactant stoichiometry problem. (My favorite.) I'm going to solve it using Method 2.

.245 L of HNO_3 x $\frac{.784\ mol\ HNO3}{1\ L\ HNO3}$ x $\frac{2\ mol\ H2O}{2\ mol\ HNO3}$ x $\frac{18\ g\ H2O}{1\ mol\ H2O}$ = 3.46 g H_2O

.541 L $Ca(OH)_2$ x $\frac{1.79\ mol\ Ca(OH)2}{1\ L\ Ca(OH)2}$ x $\frac{2\ mol\ H2O}{1\ mol\ Ca(OH)2}$ x $\frac{18\ g\ H2O}{1\ mol\ H2O}$ = 34.9 g H_2O

Because these reactions would be happening simultaneously, only the smaller amount can be true. Therefore, the answer to the question is 3.46 g H_2O.

Chapter 11:
Energy and Disorder

Introduction:

This unit will focus on the flow of energy as well as the tendency of molecules to spread out.

Definitions:

Thermodynamics: study of heat and the relationships between different forms of energy
Heat (q): a form of energy associated with molecule movement
Temperature: a measure of heat; energy flows from areas of high temperature to areas of low temperature
Endothermic processes: the flow of energy is into the system from the surroundings
Exothermic processes: the flow of energy is out from the system and into the surroundings
First Law of Thermodynamics: heat is a form of energy and obeys the law of conservation of energy
Law of Conservation of Energy: energy can't be created or destroyed, but it can change forms readily
Specific heat: the amount of heat needed to raise a certain mass of a substance by one degree Celsius
Enthalpy: heat content of a system. This cannot be measured, but we can measure the ΔH or change in the enthalpy.
Entropy: Disorder

Molar enthalpy (heat) of vaporization (ΔH_{vap}): heat required to vaporize one mole of a liquid
Molar enthalpy (heat) of fusion (ΔH_{fus}): heat required to melt one mole of a solid substance
Note: ΔH values will be positive because vaporizing a liquid and melting a solid are endothermic processes.

Phase Changes:

Melting: solid to liquid
Freezing: liquid to solid
Vaporization: liquid to gas
Condensation: gas to liquid
Sublimation: solid to gas
Deposition: gas to solid

Two Key Equations:

The one you use is determined by the information you're given.
If there's a change in temperature use: $q = m\ c\ \Delta T$
q is heat absorbed or released (J or kJ)
m is mass of the substance (grams)
c is the specific heat of the substance (usually given to you; units = J/g*K)
ΔT is the change in temperature (°C or K); you may have to calculate this

Important information:
1000 J = 1 kJ
$\Delta T = T_{final} - T_{initial}$

Other versions of the equation:
It's helpful to be able to solve for any variable within an equation. You're not always asked to solve for heat. Sometimes, you're asked to solve for the mass, the specific heat, or even the change in the temperature. It's even possible to ask you to solve for the final or the initial temperature, though I typically don't ask that of a college prep class.

$q = m\,c\,\Delta T \qquad m = \frac{q}{c\,\Delta T} \qquad c = \frac{q}{m\Delta T} \qquad \Delta T = \frac{q}{mc}$

Endothermic vs. Exothermic:
Endothermic reactions have positive q values, and exothermic reactions have negative q values. Endothermic and exothermic are tricky. It's almost counterintuitive (against what you'd normally think). Endothermic reactions like ammonium chloride in water get colder. These are endothermic because the reaction absorbs the heat. We're sensing the change in the water, so we feel it getting colder. Exothermic reactions like calcium chloride in water get warmer. These are exothermic so the reaction gives off heat, but we're sensing the change in the water, which gets warmer.

If there's a change in the state of matter (phase change) use:
$q = m\,\Delta H_{fusion/vaporization}$ or $q = mol\,\Delta H_{fusion/vaporization}$
Note: the difference depends on the units of $\Delta H_{fusion/vaporization}$

Calorimeter: something used to measure the change in heat during a chemical reaction or process

Food labels:

You might be asked to read and interpret a food label. The calories you read on there are actually kilocalories, which are 1000 of the small calories.

Heating Curves:

These will show the temperature changes of a substance that's heated up at a constant rate. They start low and slope up and to the right, flatten out, rise again, flatten out again, and rise some more.

The first sloping line at the bottom left of the heating curve represents the substance in the solid state of matter. As heat is poured into the substance, the temperature will rise until you reach the substance's melting point. There, the line will flatten out. Energy is still going in but the temperature doesn't change because that extra energy is being used to break bonds. Once the phase change to liquid is complete, the temperature will rise again until the substance reaches the boiling point. Then, it will flatten out again as the energy is used to accomplish the phase change from liquid to gas (vapor). If yet more energy is put in, the temperature will rise again, heating the gas.

Note: If the temperature changes as energy is put in, there is a change in the average kinetic energy. If the temperature does not change, there is a potential energy change.

Note: A cooling curve is the opposite. It will start up high, slope down then flatten out as the substance changes from a vapor to a liquid. Then, it will slope down again until it flattens out for the phase change from liquid to solid.

Calculations for Heating Curves:
The math involved in this part is simply $q = mc\Delta T$ anywhere the temperature is changing. In other words the parts that slope upward. The flat lines will require you to use $\Delta H_{fusion/vaporization}$ to determine the amount of energy needed to change the phase.

Phase Diagram:

This will show you which state of matter a substance will be in for certain pressure and temperature conditions.

Critical temperature: the temperature above which you could apply as much pressure as you want and still fail to liquefy the substance.

Triple point: the temperature and pressure condition of a substance where all three states of matter are in equilibrium. There are some rather strange Youtube videos of this.

How can we explain the process of evaporation?

Vapor pressure: the pressure of a vapor that is touching its liquid or solid form

According to the Kinetic Molecular Theory of gases, the more gas particles there are in a closed container, the more collisions there will be with the walls of the container, thus the pressure will be higher. Therefore, a higher temperature will give more molecules the energy needed to become gas particles, increasing the vapor pressure.

Boiling occurs when the vapor pressure of a liquid and the atmospheric pressure of the gas above it are equal. This means that if you lower the atmospheric pressure, you can get a liquid to boil at a lower temperature.

The infamous eggs on a mountain question:
The question might pop up here, but it's more likely to spring up during the gases unit. It's relevant here, so I might as well explain it here too. On top of a mountain (or at any high altitude) there is low air pressure, so the boiling point is low. When you try to boil something under these conditions, it's more difficult because the temperature at which you're cooking the substance is lower. Therefore, it takes longer to cook.

Pressure cookers are the opposite concept. They increase the pressure over a liquid, causing a delay in the boiling point. The substance ends up cooking at a higher temperature, so the cooking time is shorter.

Hess's Law:

Hess's Law states that no matter how many steps are involved in a chemical reaction, the total enthalpy change is equal to the sum of the changes for the individual steps.

Note: It takes energy to break bonds. This process is endothermic. Bond formation is exothermic. A positive ΔH value means the enthalpy change for that step is endothermic. A negative ΔH value means the enthalpy change for that step is exothermic.

Overview of solving Hess's Law problems:
I enjoy Hess's Law problems because they're like a puzzle. You're told what target equation to aim for.

Steps to solving Hess's Law problems:

Step #1: Flip the given equations until you get the reactants and products of the target equation on the correct sides.
Step #2: Multiply/Divide the given equations until you have the correct number of reactants and products as the target equation requires.
Step #3: Whatever you've done to the given equations must be done to the ΔH for that equation. For example, if you reversed the equation, switch the sign on the ΔH. If you divided the equation by 2, then cut the ΔH in half.
Step #4: Add the ΔH of the given equations to get the ΔH for the target equation.

Standard enthalpy of formation: the enthalpy change involved in forming one mole of a substance from its pure elements. Everything is at standard conditions.

Standard temperature and pressure (STP): 1 atm and 298 K.

Note: There are hundreds of charts for this. Solving these problems usually comes down to being careful while adding and watching signs.

This is just one of them: https://schoolworkhelper.net/standard-enthalpies-of-formation/

Second Law of Thermodynamics:

Entropy of the universe always increases. The same holds true for an isolated system.

Entropy (S): disorder. We can't measure it directly, but we can gauge the change in the entropy (ΔS).

Entropy Illustration: Top = neat room, low entropy; Bottom = messy room, high entropy

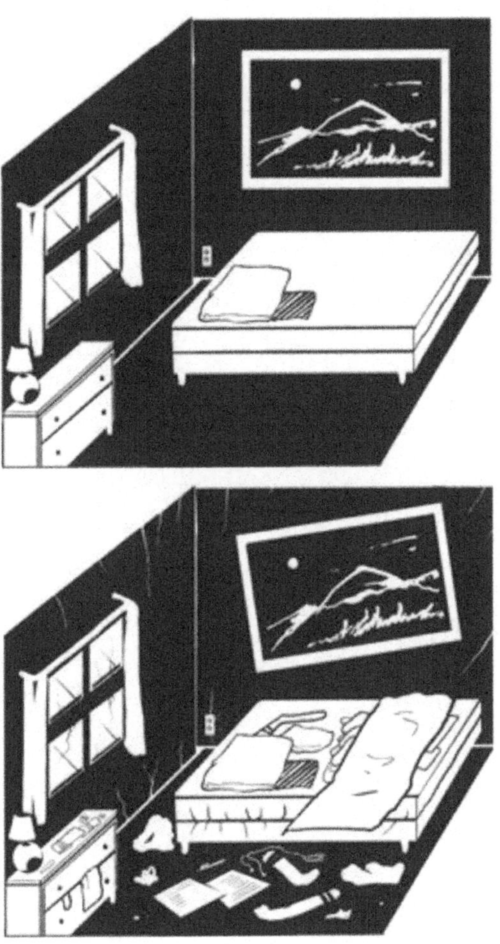

Entropy and the states of matter:
Solids have the least entropy or the most order. Liquids have more entropy than solids but less than gases. Gases have the most entropy and the least order to their molecules.

If a chemical reaction increases the number of gas particles, it will increase the entropy.

Chapter 12: Reaction Rates (Kinetics) and Equilibrium

Introduction:

Some college prep courses avoid discussing rate laws or mechanisms because few teachers want to dive into either hot mess unless they truly must. I'll discuss them briefly, but if your course doesn't require them, you might as well skip those small sections.

Key Questions about Kinetics:

What makes a chemical reaction happen? How quickly do chemical reactions take place? What can we do to make a reaction faster?

Topics: collision theory, making reactions faster, rate laws, reaction mechanisms, potential energy diagrams

Collision theory: In order for a chemical reaction to occur molecules have to collide with enough energy and the correct orientation to create the activated complex or transition state. This is the thing that will turn into products eventually.

Reasons a chemical reaction could fail are therefore, 1) not enough energy in the collision and 2) a collision with improper orientation. By orientation, we mean direction.

Puzzle analogy:
It's like fitting two puzzle pieces together. You may have the correct pieces but if you don't actually press them together, they won't form the picture you want. You could also be pressing with all your might but have them lined up incorrectly, which again results in no final picture.

What can make a chemical reaction faster?

In short, anything that can increase the number of effective collisions will make the reaction faster. There are five basic ways to ensure a higher number of effective collisions.
1. decrease volume of the reaction container
2. increase the amount of reactants
3. increase the temperature
4. increase surface area
5. add a catalyst

1. If you decrease the volume of the reaction container, the molecules are in a tighter space and bump into one another more often.
2. If you increase the amount of reactants you have, there will be more molecules available to collide with each other.
3. If you increase the temperature of the reaction, the molecules will be moving faster, causing more collisions.
4. If you increase the surface area of the reactants, you expose more molecules to the possibility of a collision. If the surface area is low, the inside molecules are protected from collision.
5. If you add a catalyst, the reaction will have a lower activation energy, which means more of the collisions that happen will form product.

Activation energy:

In the simplest terms, activation energy is the energy needed to make a reaction happen.

Sale analogy: I like to think of this in terms of the cost to buy something. If the cost to buy a particular item is high, fewer people have the money to purchase the item. This is like having a high activation energy. If the cost to buy that same item is low, more people have the money to purchase the item. This is similar to a low activation energy. Keeping with that analogy, adding a catalyst would be like putting the item on sale. By lowering the item cost, you allow more people to successfully purchase the item. By lowering the activation energy, you allow more of the collisions that happen to meet the energy requirements to successfully turn into product.

Useful Video: Search Youtube for "How to speed up chemical reactions and get a date."

What the heck are rate laws?
Confession: I hated kinetics for a very long time. It's not so bad now, but it's still not my favorite topic.

Rate laws are essentially math equations that show the relationship between the concentration of reactants and how quickly the reaction is taking place. These have to be determined experimentally, so you'll see large charts when it comes to this type of problem.

How do you write a rate law?

General form: Rate = k $[A]^x[B]^y$
The brackets just mean molar concentration, so the units are mole/L.

You need to determine the order with respect to each reactant. Let's assume we have 2 reactants, A and B. We'll hold one [B] constant and let the other [A] double and then see what happens to the rate. If the rate doubled, then you'd be able to sum it up like this. If B's constant, A increased x 2 and rate increased x 2. $2^x = 2$, $x = 1$. The order with respect to A would be 1 or first order.

You would do the same kind of calculation for B. Hold A constant, let B change. Let's say B doubled and the rate quadrupled. B increased by x 2 and rate increased x 4. $2^y = 4$, $y = 2$. The order with respect to B would be 2 or second order.

The overall reaction order would be the sum of the individual orders. In other words, add x and y. In this case, the reaction order would be 3.

Solving for k:
Pick any one of the experiments, plug in the numbers and solve. I recently found out that doing this without units is helpful. The units for k can be a pain because they're going to be determined by the order of the reactants.

Tip: It's easier to solve for the units separately. Remember mol/L is the same thing as M, so where possible, see if you can simplify the units before solving for them. Just to confuse people, sometimes the problems will have units like mol L^{-1} s^{-1}. This is actually the same thing as writing $\frac{mol}{L*s}$.

Reaction mechanisms:
Reaction mechanisms are the overall series of steps which result in the complete balanced equation. The extent to which a first-year chemistry course typically deals with this topic is to make you recognize the difference between an intermediate and a catalyst.

Intermediate: This is the product of one elementary step and the reactant in a subsequent elementary step.

Catalyst: A catalyst will speed up a chemical reaction without being consumed by that reaction. It's going to be seen in the reactants of the first step and then again in the products of the last step.

Rate determining step: This will be the slow step. Even if you change the fast step of the mechanism, it won't speed up the reaction as a whole because it's the slow step that controls how quickly the chemical reaction can take place.

Potential energy diagrams:
You will likely have to be able to identify the reactants, products, activation energy (Ea), activated complex/transition state, ΔH, and possibly the reverse activation energy (Ea'). If you throw the words potential energy diagram into any search engine, you'll come out with a few million examples of these. You may also have to be able to show where on a potential energy diagram you would see the effects of a catalyst. The answer is the activation energy. Catalysts lower activation energy, so the "hump" up to the activated complex would be smaller.

Endothermic potential energy diagram:
The reactants will be lower than the products. So, it'll start off low and swing high, indicating a high activation energy. This makes sense because it takes energy to make an endothermic reaction work.

Exothermic potential energy diagram:
The reactants will be higher than the products. The reactants will start fairly high, swing up to the activated complex, then flow down to products. There's still activation energy involved, just a smaller amount than the endothermic reactions.

Chemical Equilibrium:

Equilibrium is reached when the forward reaction and the reverse reaction happen at equal rates. That's why this topic tends to land right after the discussion of reaction rates. The forward reaction is from reactants to products. The reverse reaction is from products to reactants.

When talking about equilibrium, we're typically discussing Le Chatelier's Principle. There are literally dozens of practice sheets available online, but let's talk concept for a bit. Le Chatelier's Principle is about undoing what you just did to a chemical reaction.

Anything you do to stress the reaction will cause the reaction to shift. Your job with Le Chatelier's Principle problems will be to predict the direction of that shift.

What could cause the equilibrium position to shift?
1. changing the concentration of reactants or products
2. adding or removing heat
3. changing the size of the container/ changing the pressure

Concentration of reactant or products:
Tip: Think of it this way. If you add something, the shift is away, so the shift is to the opposite side of the reaction. If you remove something, you must replace it, so the shift is to the same side as what you just removed.

Trick - Adding or removing heat:
This one's difficult because it can differ based on which side the heat's already on. If you see a $-\Delta H$, that means the heat is a product. If you see a $+\Delta H$, heat energy is a reactant. This will make your life easier. I highly recommend actually plugging in the heat as if it was a reactant or a product. Then, you can treat it as such when it comes to the equilibrium shift.

Example 1: If heat is a reactant ($+\Delta H$) and you raise the temperature, it's the same effect as if you'd added one of the reactants. The shift would be to the right side, favoring the forward reaction.

Example 2: If heat is a product (-ΔH) and you raise the temperature, it's the same effect as if you'd caused a build-up on products. The shift would be to the left side, favoring the reverse reaction.

Note: Pure solids and liquids do not affect the equilibrium position.

Conclusion:

Much of life involves balance. Chemistry is no exception. Equilibrium is the part of chemistry that deals with maintaining balance.

Chapter 13:
Acids and Bases

Introduction:

Your textbook will likely talk a lot about the properties of acids and bases.

Category	Acids
Physical and Chemical Properties	Taste sour Conducts electricity Aq acids turn blue litmus paper red React with metals to release H_2 gas React with bases to produce salt and water contains more hydrogen ions (H^+) than hydroxide (OH^-)
Common Examples	HCl (hydrochloric acid) H_2SO_4 (sulfuric acid) H_3PO_4 (phosphoric acid) HNO_3 (nitric acid) $HClO_4$ (perchloric acid)
Category	Bases
Physical and Chemical Properties	Taste bitter Feels slippery Conducts electricity Aq bases turn red litmus paper blue React with acids to produce salt and water contains more hydroxide ions (OH^-)

	than hydrogen ions (H^+)
Common Examples	$NaOH$ (sodium hydroxide) NH_3 (ammonia) Na_2CO_3 (sodium carbonate)

Why should you care?

Acids and bases are another great area of chemistry that has real world applications.

Reason #1: If you ever need to take care of a pool, you'll likely have to balance the pH. That's why every pool store sells pH testing kits. You need the pH balanced correctly so the water is unfriendly to bacteria that would grow in your pool.

More on this here:
http://www.swimmingpool.com/maintenance/general-maintenance-and-tips/importance-balanced-pool

Reason #2: A lot of common household items have pHs you should be aware of in order to handle them safely. Cleaning products are highly basic, which is good because that means they're detrimental to the life forms growing in weird places around your house. This is also bad though because you're a lifeform too, so you need to handle cleaning products with care to avoid injury.

Reason #3: It can help you stay healthy. In short, acids corrode, so you want to limit the intake of acidic foods. This is an argument against sucking on lemons.

Article that goes into great detail on this:
http://www.nachicago.com/CHI/January-2012/What-is-pH-and-Why-Should-I-Care/

Reason #4: You can use a working knowledge of ascorbic acid to prevent fruit from browning.

Sources:
http://www.aces.edu/dept/extcomm/specialty/antibrowning.html
http://www.livestrong.com/article/496950-is-ascorbic-acid-a-preservative/

Note: Crash Course Chemistry #30 has a great summary of pH and pOH.

Here is the link to that:
https://www.youtube.com/watch?v=LS67vS10O5Y

Definitions:

Dissociates: comes apart in water to form ions
Proton: H^+ ion
Hydrogen: H^+ ion
Hydroxide: OH^- ion
Aqueous: a solution where water is the solvent
Amphoteric: something that can act as an acid or as a base. The most common example is water. It can give up a hydrogen ion, thereby acting as an acid, or it can accept a hydrogen, thereby acting as a base.
Neutralization: a chemical reaction between an acid and a base.

Hydrogen and Hydronium:

The amount of hydrogen ions (H^+) in solution determines how acidic it is. So, we typically need to track the amount of H^+ ions in solution. But H^+ doesn't stick around long in water. It immediately reacts with a water molecule (H_2O) to produce a hydronium ion (H_3O^+). This leads many people to use the terms, H^+ and H_3O^+ interchangeably.

Three Definitions of Acids and Bases:

There are three different definitions of acids and bases: Arrhenius, Brønsted-Lowry, and Lewis.

1) The Arrhenius Model (cannot explain bases like ammonia and sodium carbonate)

Arrhenius Acid: produces hydrogen ions in aqueous solution; i.e. HCl

Arrhenius Base: produces a hydroxide ion in aqueous solution; i.e. NaOH

2) The Brønsted-Lowry Model

Brønsted-Lowry Acid: hydrogen (H^+) donor

Brønsted-Lowry Base: hydrogen (H^+) acceptor

3) The Lewis Model

Lewis Acid: Electron pair acceptor

Lewis Base: Electron pair donor

Monoprotic, Diprotic, and Triprotic:

A proton is an H^+ ion.

Monoprotic: An acid that releases one H^+ ion when it dissociates (comes apart) in water is monoprotic. Hydrochloric acid (HCl) is an example of a monoprotic acid.

Diprotic: An acid that releases two H^+ ions when it dissociates in water is diprotic. Sulfuric acid (H_2SO_4) is an example of a diprotic acid.

Triprotic: An acid that releases three H^+ ions when it dissociates in water is triprotic. Phosphoric acid (H_3PO_4) is an example of a triprotic acid.

Note: Bases have the same issue. Some like, sodium hydroxide [NaOH], release 1 OH^- ion when dissociated in water. Others, like magnesium hydroxide [$Mg(OH)_2$], release 2 hydroxide ions when dissociated in water. And still others, like aluminum hydroxide [$Al(OH)_3$], release 3 hydroxide ions when dissociated in water.

Why should you care?

Depending on how challenging your teacher wants to get, you might be given solutions of strong acids like sulfuric acid, which is diprotic. That means that for every one molecule of H_2SO_4 broken down, there will be two H^+ ions released. This will affect the number of hydrogen ions you get when the acid dissociates. It will be 2x greater than the molarity of the solution.

Strength of Acids and Bases:

The strength of an acid or base typically refers to the extent to which it dissociates in water. Strong acids and bases dissociate completely. This means that if you put one HCl molecule into water, you would get it to break down into one H^+ ion and one Cl^- ion. Weak acids and bases dissociate partially.

The pH Scale:

The pH scale measures the power of the hydrogen ion you get in solution. If you have extra H^+ ions in solution, you will have an acid. If you have extra OH^- ions in solution, you will have a base. The scale itself runs from 0 to 14. The 0 side represents very strong acids. The 14 side represents very strong bases. 7 is neutral. The closer you get to 7, the weaker your acid or base will be.

Common pH's:

It's hard to predict what your teacher will want in terms of common pH's. Sometimes, you'll have to use a pH scale to determine the acidity or alkalinity (how basic) an item is, and other times, you may have to go the other way.

These sites have some of the most common acid and base data:
http://madang.ajou.ac.kr/~ydpark/lectures/chem_cntxt/acid_base/pH_common_materials.htm

http://www.nachicago.com/CHI/January-2012/What-is-pH-and-Why-Should-I-Care/
One of the nicest charts can be found here:
https://www.pmel.noaa.gov/co2/file/The+pH+scale+with+some+common+examples
To learn more about the PMEL Carbon Program, please visit their website.

Highlights:

Substance	Approximate pH	Interpretation
Battery acid	0	Very, very acidic (take-your-face-off dangerous)
Lemon juice	2	Very acidic
Black coffee	5	Acidic
Milk	7	Neutral
Seawater	8	Mildly basic
Baking soda	9	Basic
Household ammonia	11	Very basic
Household bleach	12.5	Very, very basic (take-your-face-off dangerous)
Sodium Hydroxide	14	Very, very, very basic

Calculating pH:

Here's every pH formula and the various versions that I think you will need.

$14 = pH + pOH$
$pH = 14 - pOH$
$pOH = 14 - pH$

$pH = -\log [H_3O^+]$
$[H_3O^+] = 10^{(-pH)}$
$pOH = -\log [OH^-]$
$[OH^-] = 10^{(-pOH)}$
$1.0 \times 10^{-14} M^2 = [H_3O^+] [OH^-]$
$[H_3O^+] = 1.0 \times 10^{-14} M^2 / [OH^-]$
$[OH^-] = 1.0 \times 10^{-14} M^2 / [H_3O^+]$

Note: I have centered the main formulas and left justified the ones that are reworked versions so you can distinguish between them. If you are required to memorize the pH calculation formulas, the centered ones are the ones to memorize. The others can be solved for algebraically. If Algebra's not your thing but memorizing is easy for you, learn the whole chart.

Note: $[H_3O^+]$ is the hydronium ion. You may see $[H_3O^+]$ and $[H^+]$ used interchangeably. Protons, H^+'s, do not stay around long. They immediately combine with a water molecule to make a hydronium ion.

Example: Determine the $[OH^-]$, $[H_3O^+]$, pOH and pH of a 0.045 mol/L HCl solution. Explain your answer for H^+ concentration. (**Note:** you need to use exact calculator number to get the right answer).

$[H^+]$ = .045 M because each mole of HCl will dissociate into a mol of H^+ ions
$pH = -\log [H_3O^+]$
$pH = -\log [.045] = 1.35$
$pOH = 14 - pH$
$pOH = 14 - 1.35 = 12.65$
$[OH^-] = 1.0 \times 10^{-14} M^2 / [H_3O^+]$
$[OH^-] = 1.0 \times 10^{-14} M^2 / .045 M =$
$[OH^-] = 2.22 \times 10^{-13}$ M or get 2.2×10^{-13} M.

Explanation: The $[H^+]$ is .045 M because each mole of HCl will dissociate into a mol of H^+ ions. The pH can be calculated by using the normal pH formula: pH = -log $[H^+]$. When I plug in the .045 M, the pH is 1.35. Now that I have the pH, the pOH is easy to calculate because they have to equal the total of the pH scale, which is 14. Since pH plus pOH is 14, 14 minus a pH of 1.35 will tell us that the pOH is 12.65.

There are two ways to solve for the hydroxide ion, $[OH^-]$. I chose to use the relationship between hydroxide and hydronium ions: $1.0 \times 10^{-14}\ M^2 = [H_3O^+]\ [OH^-]$. When I divide $1.0 \times 10^{-14}\ M^2$ by .045 M, I get 2.22×10^{-13} M or get 2.2×10^{-13} M.

Alternate Solution: The number 1.0×10^{-14} tends to scare chemistry students. If you prefer using antilog, you should still get very similar answers.
$[OH^-] = 10^{(-pOH)}$
$[OH^-] = 10^{(-12.65)}$
$[OH^-] = 2.24 \times 10^{-13}$ M or 2.2×10^{-13} M

Explanation for Alternate Solution: The hydroxide ion concentration can be found by taking the antilog of the negative pOH. Essentially, I need to "undo" the pOH formula. Recall that pOH was the equivalent of the negative log of the hydroxide ion concentration. When I take the antilog of -12.65, the answer becomes 2.24×10^{-13} M or 2.2×10^{-13} M.

Conclusion:

Like the gas laws, the acids and bases chapter can be fascinating in terms of its real world applications. The math might not be your favorite, but it is doable.

Conclusion: Wading through Resources

Introduction:

Too much information's almost as useless as not enough information. That's one of the biggest problems I can foresee chemistry students having. All of the information's out there, it's just overwhelming or not easy to find. The rest of the book has focused on the other skills you need to survive chemistry, but I saved the fifth skill of knowing when and how to seek extra help for last.

Resources Available to Students:

- Your teacher
- Class notes/worksheets and homework
- Your textbook/online textbook
- The internet: websites
- The internet: YouTube and other video tutorials

The single best resource at your disposal (most of the time) = your teacher

As you've glimpsed in this book, there are several ways most of these topics can be looked at. In order to succeed in the class, you need to understand your teacher and his or her method for doing each kind of problem. Everybody chooses something to emphasize. For me, it's units. I put units everywhere and ask my students to do the same. I think it makes everything easier to follow. Your teacher may or may not place the same emphasis on units. Their focal point could be something else like sig figs.

We live in a day and age where most teachers have email. The extent to which they check said email will vary widely. If your teacher is available by email, ask him or her questions that way or at least use it to set up extra help appointments.

Official requirements for extra help vary by school and school district, but in general, most teachers are available some time outside of school hours to answer questions and give extra help.

Extra help vs paid tutoring:

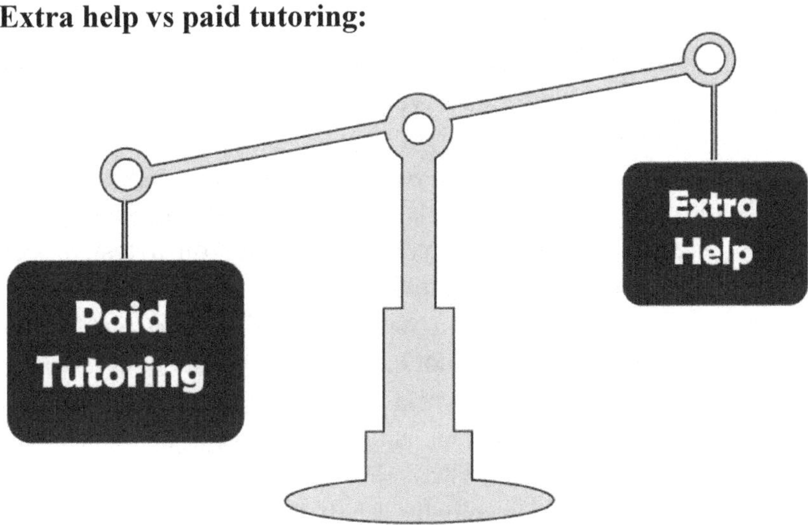

Please understand that there's a distinct difference between extra help and paid tutoring. Your teacher is your best resource because he or she will create the tests, quizzes, projects, and other assessments you'll be asked to do. But extra help is NOT tutoring.

To get the most out of extra help, attempt to do the homework problems yourself and come with specific questions.

One of the first questions you're likely to be asked is: "What do you need help with?"

If you respond with, "Everything." It's going to be a lot harder to get you the help you need because time will have to be spent determining where you got lost along the way. A forty-to-sixty-minute lesson isn't easy to squeeze into a ten-to-twenty-minute extra help session.

If you're going to seek extra help, it's always best to set up an appointment, but know that you may not be the only one there at that time, especially if it's the day of a test or quiz.

The differences:

- Paid tutors can earn upwards of $100 an hour (or more) to work with you exclusively for the timeframe. Extra help with your teacher is free.
- Your tutor can help you with homework problems or concepts covered in class, but he or she probably can't predict exactly what you'll encounter on a test or quiz since he or she didn't make it.
- Your tutoring session will likely be one to one, but an extra help session is not exclusive to you.
- A paid tutor may have a different approach to problems than your teacher. This is usually fine, but be aware that a difference could exist.
- Paid tutoring is typically a half-hour, an hour, or an hour and a half. I suppose it could stretch on longer, but generally, people start to fade out after an intense hour and a half. Extra help is whatever time you can squeeze in with the teacher before school, after school, or during lunch.

Nonpaid tutoring:

Often times, the National Honor Society students need to log certain hours of community service and tutoring counts. You might be able to set up regular appointments with a more experienced student.

Your friends in the same class or a different chemistry class may also be a good resource. I maintain that your teacher is your best bet for extra help, but friends definitely win the category of accessibility.

Class notes, worksheets, and homework:

These are excellent resources, but typically, on their own, they're not capable of giving students a thorough look at a topic. Worksheets focus on one aspect of a lesson.

Textbooks:

If I had to guess the single most underused resource, I'd say the textbook. The effectiveness of textbooks varies, as does the reading level. Most high school chemistry textbooks are fine, but they still contain an overwhelming amount of information that students are disinclined to dig out. They are great for brushing up on terms and they often have great examples and a ton of extra problems. However, there's often so much to take in that students have a hard time picking out the most important points.

Find a way that works for you. Highlight as you go, if you own a copy of the book. Most textbooks have older versions available online from resellers. If you really need to highlight something but don't own the book, consider scanning it into a computer and printing the relevant pages or getting a photocopy of a certain chapter. Take notes in some other way. Type out notes. Record yourself reading sections. Repeat sections privately to yourself. Re-type certain passages. People learn different ways.

Internet websites:

The most unpredictable award for resource goes to internet websites. These can be gold mines of information or blithering piles of falsehoods. Wikipedia is an okay starting point, but it's not the most reliable source out there. Typically, something with a .edu or .gov should be higher on the reliable side.

Besides the question of true or false, internet websites vary widely on the issue of comprehension. Some are clearly meant for college chemistry students. As you dive into a site, use your best judgement. If the terminology and language seems above your head, find a different site. If you type anything into a search engine, you're likely to get thousands of hits.

Even blogs have some great information. These are good for brushing up on a topic, but I wouldn't go quoting them for a research report.

Forums and things like Yahoo answers where anybody and everybody can answer a question are sometimes helpful and sometimes just wrong, so use these with caution.

Internet videos:

There are some excellent videos out there, but there are probably hundreds of thousands of other videos that are wrong, misleading, or bad in some other way. I've not watched all of the Crash Course Chemistry videos, but the ones I have seen are good introductions to their topics. Kahn Academy and Ben's Chem Videos also have some decent videos, though I've definitely seen way less of these, so I can't give an official endorsement either way to the channels as a whole.

When stumped on a topic, your best bet is to ask your teacher if he or she has a recommendation for a video. If they don't get back to you in time or never look at email at all, type the topic title into YouTube. See what the first five videos are and start watching them. Usually, you can tell in the first thirty seconds whether or not you'll like a video. As for gauging usefulness, check out the number of views the video has, the likes/dislikes, and skim some of the comments. Between those, you should be able to tell how helpful the video has been to other people.

How to Talk to (and Email) Adults (Specifically, Your Teacher):

- Be brief.
- Be clear.
- Be polite.
- Be you.

There are many reasons to contact your teacher(s). Sometimes, you need to inform your teacher(s) of a planned absence due to a field trip, a family vacation, or a death in the family. Sometimes, you need to ask about the details to a project or homework assignment. Check the class website first, but they're not always updated. Sometimes, you need to ask for an extension on an assignment due to extenuating circumstances. Maybe you want to know how you did on a test or need to get a grade changed because what the paper says is different than what the online system says. The scenarios are pretty much endless, but the four bullet points above are your best bet for obtaining the most favorable outcome of the situation.

By the time you're in chemistry, you're in high school. It's time for you to present your own requests. I'm not saying there aren't situations that warrant involving your parents, but they should not be your first go-to option. Any time you involve additional people, things get more complicated.

Mistakes happen. I've definitely entered grades wrong, and in those situations, I'm happy to fix the problem as swiftly as possible. If you've earned a grade, you deserve it to be reflected properly in the overall scores. That said, do not send requests to have a grade changed based on the reasoning that you "worked hard." Working hard is the expectation, and the grade earned will already reflect this. 89's do not magically turn into 90's because you're nice. If they did, the entire grading system would become arbitrary and meaningless.

When you compose an email to a teacher, you want to be clear and to the point as well as polite and respectful. Remember, if you're contacting your teacher, typically it's because you want something. People respond better to nice, polite emails. Part of being clear is stating who you are and possibly what period you're in. You may or may not get your way with the request, but you're more likely to get a favorable response with something clear, genuine, and polite. Always take a moment to double check the grammar too. It's just a good habit for corresponding with anybody in a formal setting.

Recap:

Closing Remarks:

Chemistry is challenging. Approach it with the attitude that you can do it, even if it's not going to be easy. Success hinges on that first step to surviving the course: having a helpful mindset and defining what success will look like. You may start out in "I hate this. This is stupid." mode, but you need to move beyond that to "This is hard but doable." Chemistry can be fun, but for the vast majority of students, it's simply something that needs to be conquered.

If you have any specific questions, my public email is: juliecgilbert5steps@gmail.com. I can't guarantee I'll know how to help, but I might be able to point you in the right direction for finding the answers you seek.

If you've found this book helpful, please consider reviewing it.

Thanks for reading.

Sincerely,

Julie C. Gilbert

Had enough chemistry for now?

Each book—fiction or nonfiction—is a labor of love, and it's my pleasure to share the story or knowledge with you.

Please visit my website: http://www.juliecgilbert.com/ to find a link to the current free works. It is my goal that individual ebooks be free, but if you still wish to show support there are combination books and other formats such as audiobooks and paperbacks to invest in.

Search for "Julie C. Gilbert's Special Agents" on Facebook for monthly book discussions and giveaways.

I would love to connect with you. Please reach out to me via email at: devyaschildren@gmail.com or juliecgilbert5steps@gmail.com.

Other Contacts:
Facebook: https://www.facebook.com/JulieCGilbert2013
Instagram: https://www.instagram.com/juliecgilbert_writer/
Twitter: https://twitter.com/authorgilbert
Bookbub Partner link: https://www.bookbub.com/authors/julie-c-gilbert

69222376R00089

Made in the USA
Middletown, DE
20 September 2019